STUDY ⊗

ATMOSPHERIC THERMODYNAMICS

GEOPHYSICS AND ASTROPHYSICS MONOGRAPHS

AN INTERNATIONAL SERIES OF FUNDAMENTAL TEXTBOOKS

Editor

B. M. MCCORMAC, *Lockheed Palo Alto Research Laboratory, Palo Alto, Calif., U.S.A.*

Editorial Board

R. GRANT ATHAY, *High Altitude Observatory, Boulder, Colo., U.S.A.*
W. S. BROECKER, *Lamont-Doherty Geological Observatory, Palisades, New York, U.S.A.*
P. J. COLEMAN, JR., *University of California, Los Angeles, Calif., U.S.A.*
G. T. CSANADY, *Woods Hole Oceanographic Institution, Woods Hole, Mass., U.S.A.*
D. M. HUNTEN, *University of Arizona, Tucson, Ariz., U.S.A.*
C. DE JAGER, *The Astronomical Institute, Utrecht, The Netherlands*
J. KLECZEK, *Czechoslovak Academy of Sciences, Ondřejov, Czechoslovakia*
R. LÜST, *President Max-Planck Gesellschaft für Förderung der Wissenschaften, München, F.R.G.*
R. E. MUNN, *University of Toronto, Toronto, Ont., Canada*
Z. ŠVESTKA, *The Astronomical Institute, Utrecht, The Netherlands*
G. WEILL, *Service d'Aéronomie, Verrières-le-Buisson, France*

VOLUME 6

ATMOSPHERIC THERMODYNAMICS

by

J. V. IRIBARNE
University of Toronto

and

W. L. GODSON
Atmospheric Research Directorate, Atmospheric Environment Service, Toronto

SECOND EDITION

D. REIDEL PUBLISHING COMPANY

DORDRECHT: HOLLAND/BOSTON: U.S.A.
LONDON: ENGLAND

Library of Congress Cataloging in Publication Data

Iribarne, J. V. (Julio Victor), 1916–
　Atmospheric thermodynamics.

　(Geophysics and astrophysics monographs ; v. 6)
　Bibliography: p.
　Includes index.
　　1. Atmospheric thermodynamics. I. Godson, W. L. II. Title.
III. Series.
QC880.4.T5174　　1981　　　　551.5′22　　　　　　81-10674
ISBN 90-277-1296-4　　　　　　　　　　　　　　　　AACR2
ISBN 90-277-1297-2 (pbk.)

Published by D. Reidel Publishing Company,
P.O. Box 17, 3300 AA Dordrecht, Holland.

First published 1973
Second revised edition 1981

Sold and distributed in the U.S.A. and Canada
by Kluwer Boston Inc.,
190 Old Derby Street, Hingham, MA 02043, U.S.A.

In all other countries, sold and distributed
by Kluwer Academic Publishers Group,
P.O. Box 322, 3300 AH Dordrecht, Holland.

D. Reidel Publishing Company is a member of the Kluwer Group.

All Rights Reserved
Copyright © 1973, 1981 by D. Reidel Publishing Company, Dordrecht, Holland
No part of the material protected by this copyright notice may be reproduced or
utilized in any form or by any means, electronic or mechanical,
including photocopying, recording or by any informational storage and
retrieval system, without written permission from the copyright owner

Printed in The Netherlands

TABLE OF CONTENTS

PREFACE	IX
PREFACE TO THE SECOND EDITION	XI
LIST OF SYMBOLS	XIII

CHAPTER I. REVIEW OF BASIC CONCEPTS AND SYSTEMS OF UNITS

1.1.	Systems	1
1.2.	Properties	1
1.3.	Composition and State of a System	2
1.4.	Equilibrium	2
1.5.	Temperature. Temperature Scales	3
1.6.	Systems of Units	6
1.7.	Work of Expansion	8
1.8.	Modifications and Processes. Reversibility	9
1.9.	State Variables and State Functions. Equation of State	10
1.10.	Equation of State for Gases	10
1.11.	Mixture of Ideal Gases	11
1.12.	Atmospheric Air Composition	12
	Problems	15

CHAPTER II. THE FIRST PRINCIPLE OF THERMODYNAMICS

2.1.	Internal Energy	16
2.2.	Heat	17
2.3.	The First Principle. Enthalpy	21
2.4.	Expressions of Q. Heat Capacities	22
2.5.	Calculation of Internal Energy and Enthalpy	23
2.6.	Latent Heats of Pure Substances. Kirchhoff's Equation	26
2.7.	Adiabatic Processes in Ideal Gases. Potential Temperature	28
2.8.	Polytropic Processes	31
	Problems	32

CHAPTER III. THE SECOND PRINCIPLE OF THERMODYNAMICS

3.1.	The Entropy	35
3.2.	Thermodynamic Scale of Absolute Temperature	36
3.3.	Formulations of the Second Principle	39
3.4.	Lord Kelvin's and Clausius' Statements of the Second Principle	39
3.5.	Joint Mathematical Expressions of the First and Second Principles. Thermodynamic Potentials	40
3.6.	Equilibrium Conditions and the Sense of Natural Processes	43
3.7.	Calculation of Entropy	45
3.8.	Thermodynamic Equations of State. Calculation of Internal Energy and Enthalpy	46
3.9.	Thermodynamic Functions of Ideal Gases	47
3.10.	Entropy of Mixing for Ideal Gases	48
3.11.	Difference Between Heat Capacities at Constant Pressure and at Constant Volume	49
	Problems	51

CHAPTER IV. WATER-AIR SYSTEMS

4.1.	Heterogeneous Systems	53
4.2.	Fundamental Equations for Open Systems	57
4.3.	Equations for the Heterogeneous System. Internal Equilibrium	58
4.4.	Summary of Basic Formulas for Heterogeneous Systems	59
4.5.	Number of Independent Variables	61
4.6.	Phase-Transition Equilibria for Water	62
4.7.	Thermodynamic Surface for Water Substance	64
4.8.	Clausius-Clapeyron Equation	65
4.9.	Variation of Latent Heat Along the Equilibrium Curve	68
4.10.	Water Vapor and Moist Air	69
4.11.	Humidity Variables	73
4.12.	Heat Capacities of Moist Air	76
4.13.	Moist Air Adiabats	78
4.14.	Enthalpy, Internal Energy and Entropy of Moist Air and of a Cloud	78
	Problems	85

CHAPTER V. EQUILIBRIUM WITH SMALL DROPLETS AND CRYSTALS

5.1.	Vapor Pressure of Small Droplets of a Pure Substance	87
5.2.	Vapor Pressure of Solution Droplets	90
5.3.	Sublimation and Freezing of Small Crystals	96
	Problems	97

CHAPTER VI. AEROLOGICAL DIAGRAMS

6.1.	Purpose of Aerological Diagrams and Selection of Coordinates	98
6.2.	Clapeyron Diagram	99
6.3	Tephigram	99
6.4.	Curves for Saturated Adiabatic Expansion. Relative Orientation of Fundamental Lines	102
6.5.	Emagram or Neuhoff Diagram	104
6.6.	Refsdal Diagram	106
6.7.	Pseudoadiabatic or Stüve Diagram	107
6.8.	Area Equivalence	107
6.9.	Summary of Diagrams	109
6.10.	Determination of Mixing Ratio from the Relative Humidity	110
6.11.	Area Computation and Energy Integrals	110
	Problems	115

CHAPTER VII. THERMODYNAMIC PROCESSES IN THE ATMOSPHERE

7.1.	Isobaric Cooling. Dew and Frost Points	116
7.2.	Condensation in the Atmosphere by Isobaric Cooling	120
7.3.	Adiabatic Isobaric (Isenthalpic) Processes. Equivalent and Wet-Bulb Temperatures	123
7.4.	Adiabatic Isobaric Mixing (Horizontal Mixing) Without Condensation	127
7.5.	Adiabatic Isobaric Mixing with Condensation	129
7.6.	Adiabatic Expansion in the Atmosphere	136
7.7.	Saturation of Air by Adiabatic Ascent	138
7.8.	Reversible Saturated Adiabatic Process	141
7.9.	Pseudoadiabatic Process	142
7.10.	Effect of Freezing in a Cloud	144
7.11.	Polytropic Expansion	146
7.12.	Vertical Mixing	147
7.13.	Pseudo- or Adiabatic Equivalent and Wet-Bulb Temperatures	149
7.14.	Summary of Temperature and Humidity Parameters. Conservative Properties	151
	Problems	153

CHAPTER VIII. ATMOSPHERIC STATICS

8.1.	The Geopotential Field	156
8.2.	The Hydrostatic Equation	159
8.3.	Equipotential and Isobaric Surfaces. Dynamic and Geopotential Height	160
8.4.	Thermal Gradients	163

8.5.	Constant-Lapse-Rate Atmospheres	163
8.6.	Atmosphere of Homogeneous Density	164
8.7.	Dry-Adiabatic Atmosphere	165
8.8.	Isothermal Atmosphere	166
8.9.	Standard Atmosphere	166
8.10.	Altimeter	168
8.11.	Integration of the Hydrostatic Equation	171
	Problems	174

CHAPTER IX. VERTICAL STABILITY

9.1.	The Parcel Method	177
9.2.	Stability Criteria	178
9.3.	Lapse Rates for Atmospheric Ascents	180
9.4.	The Lapse Rates of the Parcel and of the Environment	183
9.5.	Stability Criteria for Adiabatic Processes	185
9.6.	Conditional Instability	188
9.7.	Oscillations in a Stable Layer	191
9.8.	The Layer Method for Analyzing Stability	192
9.9.	Entrainment	195
9.10.	Potential or Convective Instability	197
9.11.	Processes Producing Stability Changes for Dry Air	201
9.12.	Stability Parameters of Saturated and Unsaturated Air, and Their Time Changes	208
9.13.	Radiative Processes and Their Thermodynamic Consequences	217
9.14.	Maximum Rate of Precipitation	224
9.15.	Internal and Potential Energy in the Atmosphere	227
9.16.	Internal and Potential Energy of a Layer with Constant Lapse Rate	230
9.17.	Margules' Calculations on Overturning of Air Masses	231
9.18.	Transformations of a Layer with Constant Lapse Rate	233
9.19.	The Available Potential Energy	235
	Problems	240

APPENDIX I: Table of Physical constants   245

BIBLIOGRAPHY   249

ANSWERS TO PROBLEMS   252

INDEX   257

PREFACE

The thermodynamics of the atmosphere is the subject of several chapters in most textbooks on dynamic meteorology, but there is no work in English to give the subject a specific and more extensive treatment. In writing the present textbook, we have tried to fill this rather remarkable gap in the literature related to atmospheric sciences. Our aim has been to provide students of meteorology with a book that can play a role similar to the textbooks on chemical thermodynamics for the chemists. This implies a previous knowledge of general thermodynamics, such as students acquire in general physics courses; therefore, although the basic principles are reviewed (in the first four chapters), they are only briefly discussed, and emphasis is laid on those topics that will be useful in later chapters, through their application to atmospheric problems. No attempt has been made to introduce the thermodynamics of irreversible processes; on the other hand, consideration of heterogeneous and open homogeneous systems permits a rigorous formulation of the thermodynamic functions of clouds (exclusive of any consideration of microphysical effects) and a better understanding of the approximations usually implicit in practical applications.

The remaining two-thirds of the book deal with problems which are typically meteorological in nature. First, the most widely-used aerological diagrams are discussed in Chapter V; these play a vital role in the practice of meteorology, and in its exposition as well (as later chapters will testify). Chapter VI presents an analysis of a number of significant atmospheric processes which are basically thermodynamic in nature. In most of these processes, changes of phase of water substance play a vital role – such as, for example, the formation of fog, clouds and precipitation. One rather novel feature of this chapter is the extensive treatment of aircraft condensation trails, a topic of considerable environmental concern in recent years.

Chapter VII deals with atmospheric statics – the relations between various thermodynamic parameters in a vertical column. In the final (and longest) chapter will be found analyses of those atmospheric phenomena which require consideration of both thermodynamic and non-thermodynamic processes (in the latter category can be listed vertical and horizontal motions and radiation). The extensive treatment of these topics contains considerable new material, which will be found especially helpful to professional meteorologists by reason of the emphasis on changes with time of weather-significant thermodynamic parameters.

The book has grown out of courses given by both of us to students working for a degree in meteorology, at the universities of Toronto and of Buenos Aires. Most of these students subsequently embarked on careers in the atmospheric sciences – some in academic areas and many in professional areas, including research as well as forecasting. As a consequence, the courses taught, from which the present text originates, emphasized equally the fundamental topic and practical aspects of the subject. It can therefore be expected to be of interest to engineers dealing with atmospheric problems, to research scientists dealing with planetary atmospheres and to all concerned with atmospheric behavior – either initially, as students, or subsequently, in various diverse occupations.

PREFACE TO THE SECOND EDITION

It is now eight years since we commented that we had produced a text book on atmospheric thermodynamics to fill a rather remarkable gap in the literature related to the atmospheric sciences – namely, a general-purpose concise text in English on thermodynamics applied to the atmosphere. We have been extremely pleased with the wide acceptance of our text, and have realized that it would be worthwhile improving on this contribution to atmospheric science teaching and research by producing a revised edition. In so doing, we have taken advantage of many helpful comments made by students and colleagues.

The general subject of atmospheric thermodynamics must be regarded as extremely stable, so that there were few new concepts that we felt deserved inclusion in this edition. However, there were some topics that we considered required a more detailed treatment; moreover, it was decided that the thermodynamic aspects of cloud microphysics should now be incorporated. Thus a new chapter on equilibrium with small droplets and crystals was added, more attention was given to polytropic processes (Ch. 2, Section 8 and Ch. 7, Section 11), a section was included on the variation of latent heat along the equilibrium curve (Ch. 4, Section 9) and the last section on available potential energy was expanded; minor corrections and improvements are also to be found in many places throughout the text. This was complemented by a list of Symbols, a rather extensive Bibliography and new additional problems; for some of the latter, we acknowledge use of the WMO publication by Laikhtman *et al.* (see complete reference in Bibliography, Section 6; their problems 2-12, 24, 33, 39, 5-17 and a 'sample problem' – p. 38 – were used as such or modified for our problems VII-12, VIII-5, IX-9, 10, 11, 12).

While we were working on this revised edition, we were saddened to learn of the death, on 31 August 1980, of Prof. Jacques van Mieghem. Prof. van Mieghem was a true pioneer in atmospheric dynamics and thermodynamics, and the textbook on atmospheric thermodynamics which he co-authored (in French) is a model of elegance and precision. As a tribute to his contributions to atmospheric science, we would like to dedicate this edition to his memory.

In closing, we would like to express our thanks to all who have helped with the two editions and to express our particular gratitude to our publisher, D. Reidel, for excellence in both printing and publishing.

LIST OF SYMBOLS

Roman Letters

a	Work performed on the system by external forces, per unit mass.
A	Work performed on the system by external forces, per mole or total. Available potential energy.
c	Specific heat capacity. Number of components.
C	Molar heat capacity.
d	Exact differential.
D	Virtual differential.
e	Water vapor pressure.
f	Specific Helmholtz function. Correction coefficient.
F	Helmholtz function, molar or total.
g	Specific Gibbs function. Gravity.
G	Gibbs function, molar or total.
h	Specific enthalpy.
H	Enthalpy, molar or total.
k	Compressibility coefficient.
K	Kinetic energy.
l	Specific heat of phase change.
L	Molar heat of phase change.
m	Mass.
M	Molecular weight.
n	Number of moles. Polytropic exponent.
N	Molar fraction.
p	Pressure.
P	Potential energy. Rate of precipitation.
q	Heat received by the system, per unit mass. Specific humidity.
Q	Heat received by the system, molar or total.
r	Mixing ratio.
R	Specific gas constant.
R^*	Universal (molar) gas constant.
s	Specific entropy.
S	Entropy, molar or total.
$\mathscr{S}$	Surface.
t	Time. Temperature, on Celsius scale.

T	Absolute temperature.
u	Specific internal energy.
U	Internal energy, molar or total. Relative humidity.
v	Specific volume.
V	Volume, molar or total.
$\mathcal{V}$	Volume of a drop.
z	Any specific property, derived from Z. Height.
Z	Any extensive property.

Greek Letters

α	Thermal coefficient.
β	Lapse rate, referred to actual height.
γ	Lapse rate, referred to geopotential.
δ	Non-exact differential. Geometric (as opposed to process) differential.
Δ	Finite difference.
ε	Ratio of molecular weights of water and dry air.
η	Ratio of heat capacity at constant pressure to heat capacity at constant volume.
θ	Potential temperature.
$\varkappa$	Ratio of gas constant to molar heat capacity at constant pressure.
λ	Rate of temperature change dT/dt.
μ	Chemical potential.
ν	Frequency. Variance of a system.
π	Length ratio of circumference to diameter.
ϱ	Density.
σ	Surface tension. Interfacial tension.
Σ	Area on a diagram.
τ	Period.
φ	Latitude. Number of phases.
ϕ	Geopotential.
ω	Angular velocity.

Subscripts

a	Adiatic (as in T_{aw}).
c	Condensed phase. Critical.
d	Dry air. Dry adiabatic. Dew point (in T_d).
e	Equivalent (as in T_e).
f	Fusion (in l_f). Final. Frost point (in T_f).
g	Gas phase.
i	Ice. Initial. Referred to ice (in U_i). Saturation with respect to ice (as in e_i). Isobaric (as in T_{iw}).

l	Liquid.
m	At constant composition (when using mass units). Moist.
n	At constant composition (when using number of moles).
p	At constant pressure.
s	Sublimation (in l_s). Solid. Saturation. Surface.
t	Triple point. Total.
v	At constant volume. Vaporization. Water vapor. Virtual (as in T_v).
w	Water. Referred to water (in U_w). Saturation with respect to water (as in e_w). Wet bulb (as in T_w).
Bar	Average (as in $\bar{T}$). Partial molar or specific property (as in $\bar{G}_v, \bar{g}_v$).
Prime	Parcel (as in T').

CHAPTER I

REVIEW OF BASIC CONCEPTS AND SYSTEMS OF UNITS

1.1. Systems

Every portion of matter, the study of whose properties, interaction with other bodies, and evolution is undertaken from a thermodynamical point of view, is called a *system*. Once a system is defined, all the material environment with which it may eventually interact is called its *surroundings*.

Systems may be *open* or *closed*, depending on whether they do or do not exchange matter with their surroundings. A closed system is said to be *isolated* if it obeys the condition of not exchanging any kind of energy with its surroundings.

The systems in Atmospheric Thermodynamics will be portions of air undergoing transformations in the atmosphere. They are obviously open systems. They will be treated, however, as closed systems for the sake of simplicity; this is legitimate insofar as we may consider volumes large enough to neglect the mixture of the external layers with the surroundings. Or we may consider, which is equivalent, small portions typical of a much larger mass in which they are imbedded; provided we are taking a constant mass for our portion (e.g. the unit mass), any exchange with the surroundings will not affect our system, as the surroundings have the same properties as itself. This approximation is good for many purposes but it breaks down when the whole large air mass considered becomes modified by exchanges with the surroundings. This may happen, due to turbulent mixing, in convective processes that we shall consider later on; this is made visible, for instance, when ascending turrets from a cumulus cloud become thinner throughout their volume and finally disappear by evaporation of the water droplets.

1.2. Properties

The complete description of a system at a certain instant is given by that of its properties, that is by the values of all the physical variables that express those properties. For a closed system, it is understood that the mass, as well as the chemical composition, define the system itself; the rest of the properties define its *state*.

Properties are referred to as *extensive*, if they depend on the mass, or as *intensive*, if they do not. Intensive properties can be defined for every point of the system; specific properties (extensive properties referred to unit mass or unit volume) among others, are intensive properties. We shall use, with some exceptions, capital letters for extensive properties and for specific properties when they are referred to one mole:

V (volume), U (internal energy), etc.; and small letters for specific properties referred to an universal unit of mass: v (specific volume), u (specific internal energy), etc. The same criterion will be followed with the work and the heat received by the system from external sources: A, a and Q, q, respectively. m for mass and T for temperature will be exceptions to this convention.

1.3. Composition and State of a System

If every intensive variable has the same value for every point of the system, the system is said to be *homogeneous*. Taking different portions of a homogeneous system, their values for any extensive property Z will be proportional to their masses:

$$Z = mz$$

where z is the corresponding specific property.

If there are several portions or sets of portions, each of them homogeneous, but different from one another, each homogeneous portion or set of portions is called a *phase* of the system, and the system is said to be *heterogeneous*.

For a heterogeneous system

$$Z = \sum_\alpha m_\alpha z_\alpha$$

where the sum is extended over all the phases.

We shall not consider, unless otherwise stated, certain properties that depend on the shape or extension of the separation surfaces between phases, or 'interfaces'. Such is the case, for instance, of the vapor pressure of small droplets, which depends on their curvature or, given a certain mass, on its state of subdivision; this is an important subject in microphysics of clouds.

It may also happen that the values of intensive properties change in a continuous way from one point to another. In that case the system is said to be *inhomogeneous*. The atmosphere, if considered in an appreciable thickness, is an example, as the pressure varies continuously with height.

1.4. Equilibrium

The state of a system placed in a given environment may or may not remain constant with time. If it varies, we say that the system is not in equilibrium. Independence of time is therefore a necessary condition in defining equilibrium; but it is not sufficient.

Stationary states have properties which by definition are independent of time. Such would be the case, for instance, of an electrical resistance losing to its surroundings an amount of heat per unit time equivalent to the electrical work received from an external source. We do not include these cases, however, in the definition of equilibrium. They may be excluded by making the previous criterion more stringent: the constancy of properties with time should hold for every portion of the system, even if we isolate it from the rest of the system and from the surroundings.

The criterion would provide a definition including certain states of *unstable* and of *metastable* equilibria. An example of unstable equilibrium is that which may exist between small droplets and their vapor, when this is kept at a constant pressure. It may be shown that if a droplet either grows slightly by condensation or slightly reduces its size by evaporation, it must continue doing so, thereby getting farther and farther away from equilibrium. A small fluctuation of the vapor pressure around the droplet could thus be enough to destroy the state of equilibrium in this case.

Examples of metastable equilibria are supercooled water in mechanical and thermal equilibrium (see Section 5) with its surroundings or a mixture of hydrogen and oxygen in similar conditions, at room temperature. In the first example, the freezing of a very small portion of water or the introduction of a small ice crystal will lead to the freezing of the whole mass. In the second one, a spark or the presence of a small amount of catalyst will be enough to cause an explosive chemical reaction throughout the system. These two systems, therefore, were in equilibrium with respect to small changes in temperature, pressure, etc. but not with respect to freezing in one case and to chemical reaction in the other. They did not change before our perturbations because of reasons which escape a purely thermodynamic consideration and which should be treated kinetically; for the molecular process to occur, an energy barrier had to be overcome. The initial ice crystal or the catalyst provide paths through which this barrier is lowered enough to allow a faster process.

We might exclude unstable and metastable cases from the definition, if we add the condition that if means are found to cause a small variation in the system, this will not lead to a general change in its properties. If this is true for whatever changes we can imagine we have a *stable* or *true thermodynamic equilibrium*. This is better treated with the help of the Second Principle, which, by considering imaginary or 'virtual' displacements of the values of the variables, provides a suitable rigorous criterion to test an equilibrium state (cf. Chapter III).

The lack of equilibrium can manifest itself in mechanical changes, in chemical reactions or changes of physical state, or in changes in the thermal state. In that sense we speak of mechanical, chemical, or thermal equilibrium. The pressure characterizes the mechanical equilibrium. Thermal equilibrium will be considered in the following section, and chemical equilibrium in Chapter IV.

1.5. Temperature. Temperature Scales

Experience shows that the thermal state of a system may be influenced by the proximity of, or contact with, external bodies. In that case we speak of *diathermic* walls separating the system from these bodies. Certain walls or enclosures prevent this influence; they are called *adiabatic*.

If an adiabatic enclosure contains two bodies in contact or separated by a diathermic wall, their properties will in general change towards a final state, reached after a long enough period of time, in which the properties remain constant. They are then said to be in thermal equilibrium. It is a fact of experience that if a body A is in thermal equilibrium with a body B, and B is in its turn in thermal equilibrium with C, A and C are also in thermal equilibrium (the transitive property, sometimes called the 'zeroth principle' of Thermodynamics).

All the bodies that are in thermal equilibrium with a chosen reference body in a well defined state have, for that very reason, a common property. It is said that they have the same *temperature*. The reference body may be called a *thermometer*. In order to assign a number to that property for each different thermal state, it is necessary to define a *temperature scale*. This is done by choosing a thermometric substance and a thermometric property X of this substance which bears a one-to-one relation to its possible thermal states. The use of the thermometer made out of this substance and the measurement of the property X permits the specification of the thermal states of systems in thermal equilibrium with the thermometer by values given by an arbitrary relation, such as

$$T = cX \qquad (1)$$

or

$$t = aX + b. \qquad (2)$$

The thermometer must be much smaller than the system, so that by bringing it into thermal equilibrium with the system, the latter is not disturbed. We may in this way define *empirical scales* of temperature. Table I-1 gives the most usual thermometric substances and properties.

TABLE I-1

Empirical scales of temperature

Thermometric substance	Thermometric property X
Gas, at constant volume	Pressure
Gas, at constant pressure	Specific volume
Thermocouple, at constant pressure and tension	Electromotive force
Pt wire, at constant pressure and tension	Electrical resistance
Hg, at constant pressure	Specific volume

The use of a scale defined by Equation (2) requires the choice of two well-defined thermal states as 'fixed points', in order to determine the constants a and b. These had conventionally been chosen to be the equilibrium (air-saturated water)-ice (assigned value $t = 0$) and water-water vapor (assigned value $t = 100$), both at one atmosphere pressure. Equation (1) requires only one fixed point to determine the constant c, and this is chosen now as the triple point of water, viz. the thermal state in which the equilibrium ice-water-water vapor exists.

The different empirical scales do not coincide and do not bear any simple relation with each other. However, the gas thermometers (first two examples in Table I-1) can be used to define a more general scale; for instance, taking the first case, pressures can be measured for a given volume and decreasing mass of gas. As the mass tends to 0, the pressures tend to 0, but the ratio of pressures for two thermal states tend to a finite value. Equation (1), applied to any state and to the triple point, would give $T = T_t(p/p_1)$; we now modify the definition so that

$$T = T_t \lim_{p \to 0} \frac{p}{p_t} \tag{3}$$

where the subscript t refers to the triple point as standard state. The temperature thus defined is independent of the nature of the gas. This universal scale is the *absolute temperature scale of ideal gases*. The choice of T_t is arbitrary; in order to preserve the values assigned to the fixed points in the previous scales, it is set to $T_t = 273.16$ K. Here the symbol K stands for *kelvin*, the temperature unit in this scale (referred to as *degree Kelvin* prior to 1967).

The two fixed points mentioned above in connection with formula (2) have the values $T_0 = 273.15$ K and $T_{100} = 373.15$ K. Another scale, the Celsius temperature, is now defined by reference to the previous one through the formula

$$t = T - 273.15 \tag{4}$$

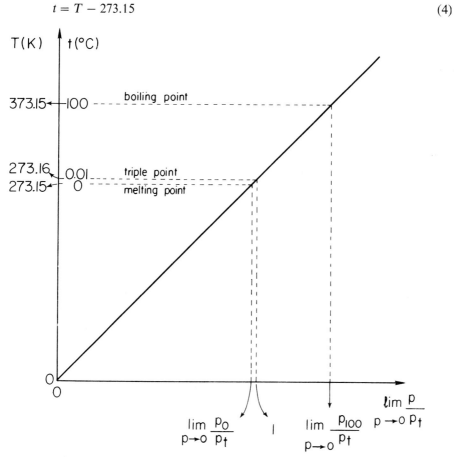

Fig. I-1. Temperature scales.

so that the two fixed points have in this scale the values $t_0=0\,°C$, $t_{100}=100\,°C$. The symbol °C stands for *degrees Celsius*. The relation between both scales is represented schematically in Figure I-1.

The absolute gas scale is defined for the range of thermal states in which gases can exist. This goes down to 1 K (He at low pressure). The Second Principle allows the introduction of a new scale, independent of the nature of the thermodynamic system used as a thermometer, which extends to any possible thermal states and coincides with the gas scale in all its range of validity. This is called the *absolute thermodynamic* or *Kelvin* scale of temperature (Chapter III, Section 2).

The Fahrenheit scale, t_F (in degrees Fahrenheit, °F), is defined by

$$t_F = \tfrac{9}{5}t + 32.$$

1.6. Systems of Units

The current systems of units for mechanical quantities are based on the choice of particular units for three fundamental quantities: length, mass, and time.

The choice of the 'MKS system' is: *metre* (m), *kilogramme* (kg) and *second* (s). The metre is defined as 1 650 763.73 wavelengths in vacuum of the orange-red line of the spectrum of krypton-86. The kilogramme is the mass of a standard body (kilogramme prototype) of Pt-Ir alloy kept by the International Bureau of Weights and Measures at Paris. The second is the duration of 9 192 631 770 cycles of the radiation associated with a specified transition of Cs^{133}. The International System of Units ('SI-Units'), established by the Conférence Générale des Poids et Mesures in 1960, adopts the MKS mechanical units and the kelvin for thermodynamic temperature.

The 'cgs system', which has a long tradition of application in Physics, uses *centimetre* (cm), *gramme* (g) and *second*, the two first being the corresponding submultiples of the previous units for length and mass.

Units for the other mechanical quantities derive from the basic ones. Some of them receive special denominations. A selection is given in Table I-2.

A third system of units, whose use was recommended at one time (International Meteorological Conference, 1911) but is now of historical significance only, is the mts, with metre, tonne (t) (1 t = 1000 kg) and second as basic units. The pressure unit in this system is the centibar (cbar): 1 cbar = 10^4 μbar.

Other units, not recommended in these systems, find frequent use and must also be mentioned. Thus we have for pressure:

    Bar (bar):   1 bar = 10^6 μbar = 10^5 Pa
    Millibar (mb):   1 mb = 10^3 μbar = 10^2 Pa
    Torricelli or mm Hg (torr):   1 torr = 133.322 Pa
    Atmosphere (atm):   1 atm = 1.013 25 bar = 760 torr =
                                             = 1.013 25 × 10^5 Pa
    Pounds per square inch (p.s.i.): 1 p.s.i. = 6894.76 Pa

The equivalence from atmosphere to bar is now the definition of the former, chosen

TABLE I-2

Derived mechanical units

Physical quantity	Unit	
	SI(MKS System)	cgs-system
Acceleration	$m\ s^{-2}$	$cm\ s^{-2}$
Density	$kg\ m^{-3}$	$g\ cm^{-3}$
Force	newton (N) $N = kg\ m\ s^{-2}$	dyne (dyn) $dyn = g\ cm\ s^{-2}$
Pressure	pascal (Pa) $Pa = N\ m^{-2} = kg\ m^{-1}\ s^{-2}$	microbar or barye (μbar) $\mu bar = dyn\ cm^{-2} = g\ cm^{-1}\ s^{-2}$
Energy	joule (J) $J = N\ m = kg\ m^2\ s^{-2}$	erg (erg) $erg = dyn\ cm = g\ cm^2\ s^{-2}$
Specific energy	$J\ kg^{-1} = m^2\ s^{-2}$	$erg\ g^{-1} = cm^2\ s^{-2}$

in such a way as to be consistent with the original definition in terms of the pressure exerted by 760 mm of mercury at 0 °C and standard gravity $g_0 = 9.80665\ m\ s^{-2}$.

The millibar is the unit which has been used most commonly in Meteorology. The listing of pressure equivalents shows that a convenient true SI unit would be the hectopascal (hPa), identically equal to a millibar. This is the pressure unit being adopted by ICAO (International Civil Aviation Organisation), effective from 26 November 1981.

Historically, an important unit for energy has been the *calorie* or *gramme calorie* (cal). Classically, it was defined as the heat necessary to raise by 1 K the temperature of one gramme of water at 15°C (cf. Chapter II, Section 2); $1\ cal_{15} = 4.1855$ J. Sometimes its multiple *kilocalorie* or *kilogramme calorie* (kcal or Cal) is used. Other definitions giving slightly different equivalences are:

International Steam Table calorie: (IT cal) 1 IT cal = 4.1868 J
Thermochemical calorie (TC cal): 1 TC cal = 4.1840 J.

The international calorie was introduced in Engineering, in connection with the properties of water substance. The thermochemical calorie was agreed upon by physical chemists, and defined exactly by the previous equivalence. Meteorologists have generally used the IT cal. However, the fundamental SI energy unit is the joule, and scientists are encouraged to avoid the use of calorie (or, as a minimum, to quote the conversion factor to joules whenever calories have had to be employed). Since the meteorological literature, and even textbooks in this area, still prefer to maintain the 'abandoned' unit (calorie), we have adopted a rather ambivalent stance in this book, and have frequently employed the symbol *cal* to denote an energy unit of 4.1868 J.

Finally, we must mention a rational chemical mass unit: the *mole* (mol). It is defined as the amount of substance of a system which contains as many elementary units (molecules, atoms, ions or electrons, as the case may be) as there are C atoms in exactly 12 g of C^{12}. It is equal to the formula mass taken in g. This 'unified definition' is slightly different from the older 'physical scale' which assigned the value 16 (exactly) to the atomic mass of O^{16} and the 'chemical scale', which assigned the same value to the natural mixture of isotopes of O.

1.7. Work of Expansion

If a system is not in mechanical equilibrium with its surroundings, it will expand or contract. Let us assume that S is the surface of our system, which expands infinitesi-

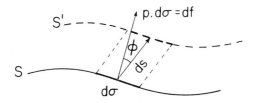

Fig. I-2. Work of expansion.

mally to S' in the direction ds (Figure I-2). The element of surface $d\sigma$ has performed against the external pressure p the work

$$(dW)_{d\sigma} = p \, d\sigma \, ds \cos\varphi = p(dV)_{d\sigma} \tag{5}$$

where $(dV)_{d\sigma}$ is the volume element given by the cylinder swept by $d\sigma$. If the pressure exerted by the surroundings over the system is constant over all its surface S, we can integrate for the whole system to

$$dW = p \int (dV)_{d\sigma} = p \, dV \tag{6}$$

dV being the whole change in volume.

For a finite expansion,

$$W = \int_i^f p \, dV \tag{7}$$

where i, f stand for initial and final states. And for a cycle (see Figure I-3):

$$W = \oint p \, dV = \left(\int_a^b p \, dV \right)_1 - \left(\int_a^b p \, dV \right)_2 \tag{8}$$

positive if described in the clockwise sense. This is the area enclosed by the trajectory in the graph.

The work of expansion is the only kind of work that we shall consider in our atmospheric systems.

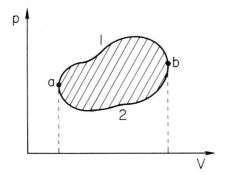

Fig. I-3. Work of expansion in a cycle.

1.8. Modifications and Processes. Reversibility

We shall call a *modification* or a *change* in a given system any difference produced in its state, independent of what the intermediate stages have been. The change is therefore entirely defined by the initial and final states.

By *process* we shall understand the whole series of stages through which the system passes when it undergoes a change.

Thermodynamics gives particular consideration to a very special, idealized type of

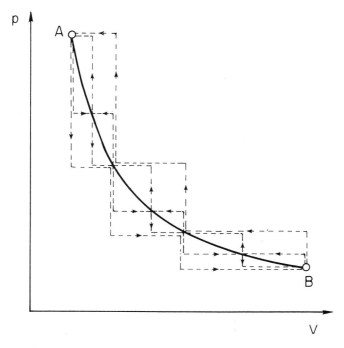

Fig. I-4. Reversible process as a limit of irreversible processes.

processes: those along which, at any moment, conditions differ from equilibrium by an infinitesimal in the values of the state variables. These are called *reversible* or *quasi-static* processes*. As an example, we may consider the isothermal expansion or compression of a gas (Figure I-4). The curve AB represents the reversible process, which appears as a common limit to the expansion or the compression as they might be performed through real processes consisting in increasing or decreasing the pressure by finite differences (zigzag trajectories). In this graph p is the external pressure exerted over the gas. The internal pressure is only defined for equilibrium states over the continuous curve AB, and in that case is equal to the external pressure.

1.9. State Variables and State Functions. Equation of State

Of all the physical variables that describe the state of a system, only some are independent. For homogeneous systems of constant composition (no chemical reactions), if we do not count the mass (which may be considered as a part of the definition of the system), only two variables are independent. These can be chosen among many properties; it is customary to choose them among pressure, volume and temperature, as readily measured properties. They are called *state* variables. All the other properties will depend on the state defined by the two independent variables, and are therefore called *state functions*. Between state variables and state functions there is no difference other than the custom of choosing the independent variables among the former.**

The equation for homogeneous systems

$$f(p, V, T) = 0$$

relating the three variables, p, V, T (two of them independent) is called the equation of state.

1.10. Equation of State for Gases

The equation of state that defines the ideal behavior for gases is

$$pV = nR^*T = mR^*T/M = mRT \qquad (9)$$

where $n=$ number of moles, $M=$ molecular weight, R^* is the universal gas constant and $R = R^*/M$ is the specific gas constant.

$$R^* = 8.3143 \text{ J mol}^{-1}\text{K}^{-1}$$
$$= 1.986 \text{ cal mol}^{-1}\text{K}^{-1}.$$

---

* These two terms are not always used as synonymous.
** It should be remarked, however, that the *thermodynamic* functions (internal energy, entropy and derived functions) differ in one fundamental aspect from the others in that their definitions contain an arbitrary additive constant. They are not really point (state) functions, but functions of pairs of points (pairs of states) (cf. following chapters).

The equation can also be written (introducing the density, ϱ) as

$$pv = \frac{p}{\varrho} = RT. \tag{10}$$

The behavior of real gases can be described by a number of empirical or semi-empirical equations of state. We shall only mention the equation of van der Waals:

$$\left(p + \frac{a}{V^2}\right)(V - b) = R^*T \quad \text{(1 mol)} \tag{11}$$

where a and b are specific constants for each gas, and the equation of Kammerlingh-Onnes, which allows a better approximation to experimental data by representing pV as a power series in p:

$$\begin{aligned} pV &= A + Bp + Cp^2 + Dp^3 + \ldots \\ &= A(1 + B'p + C'p^2 + D'p^3 + \ldots) \quad \text{(1 mol)}. \end{aligned} \tag{12}$$

$A, B, C, \ldots$ are called the virial coefficients and are functions of the temperature. The first virial coefficient is $A = R^*T$ for all gases, as it is the value of pV for $p \to 0$. The other coefficients are specific for each gas. For instance, we have for N_2 the values indicated in Table I-3.

TABLE I-3

Virial coefficients for N_2

$T(K)$	$B'(10^{-3}\,\text{atm}^{-1})$	$C'(10^{-6}\,\text{atm}^{-2})$	$D'(10^{-9}\,\text{atm}^{-3})$	pV/R^*T ($p=1$ atm)
200	-2.125	-0.0801	$+57.27$	0.9979
300	-0.183	$+2.08$	$+2.98$	0.9998

It may be seen that for pressures up to 1 atm and in this range of temperature, only the first correction term (second virial coefficient) needs to be considered.

1.11. Mixture of Ideal Gases

In a mixture of gases, the partial pressure p_i of the ith gas is defined as the pressure that it would have if the same mass existed alone at the same temperature and occupying the same volume. Similarly, the partial volume V_i is the volume that the same mass of ith gas would occupy, existing alone at the same pressure and temperature*. Dalton's law for a mixture of ideal gases may be expressed by

---

* This definition of partial volume, valid for ideal gases, would not be the proper one for real gases. The general rigorous definition of partial extensive properties, including the volume (e.g., the *partial molar* volume and the *partial specific* volume), will be considered in Chapter IV.

$$p = \sum p_i \tag{13}$$

where p is the total pressure and the sum is extended over all gases in the mixture, or by

$$V = \sum V_i \tag{14}$$

where V is the total volume.

For each gas

$$p_i V = n_i R^* T = m_i R_i T;$$

$$M_i = \frac{m_i}{n_i} = \frac{R^*}{R_i}. \tag{15}$$

Applying Dalton's law, we have

$$pV = (\sum n_i) R^* T = (\sum m_i R_i) T \tag{16}$$

where $\sum n_i = n$ is the total number of moles, and if we define a mean specific gas constant $\bar{R}$ by

$$\bar{R} = \frac{\sum m_i R_i}{m}, \tag{17}$$

the equation of state for the mixture has thus the same expressions as for a pure gas:

$$pV = nR^*T = m\bar{R}T. \tag{18}$$

The mean molecular weight of the mixture $\bar{M}$ is defined by

$$\bar{M} = \frac{\sum n_i M_i}{n} = \frac{m}{n} \tag{19}$$

so that

$$\bar{M} = R^*/\bar{R}. \tag{20}$$

The molar fraction of each gas $N_i = n_i/n$ will be

$$N_i = p_i/p = V_i/V. \tag{21}$$

Since the introduction of the mole as a fundamental unit in 1971, this ratio is often referred to as the *mole fraction*.

1.12. Atmospheric Air Composition

We may consider the atmospheric air as composed of:
(1) a mixture of gases, to be described below;
(2) water substance in any of its three physical states; and
(3) solid or liquid particles of very small size.

Water substance is a very important component for the processes in the atmosphere.

Its proportion is very variable, and we shall postpone its consideration, as well as that of its changes of state producing clouds of water droplets or of ice particles.

The solid and liquid particles in suspension (other than that of water substance) constitute what is called the atmospheric aerosol. Its study is very important for atmospheric chemistry, cloud and precipitation physics, and for atmospheric radiation and optics. It is not significant for atmospheric thermodynamics, and we shall disregard it.

The mixture of gases mentioned in the first place is what we shall call *dry air*. The four main components are listed in Table I-4; the minor components in Table I-5.

Except for CO_2 and some of the minor components, the composition of the air is remarkably constant up to a height of the order of 100 km, indicating that mixing processes in the atmosphere are highly efficient. The CO_2 has a variable concentration near the ground, where it is affected by fires, by industrial activities, by photosynthesis and by the exchange with the oceans, which constitute a large reservoir of the dissolved gas. Above surface layers, however, its proportion is also approximately constant, and can be taken as 0.03% by volume. This proportion is slowly increasing with time, and is expected to reach 0.04% by about 1990.

It may be seen that the two main gases constitute more than 99% in volume of the air; if we add Ar we reach 99.97% and if we also consider the CO_2, the rest of the

TABLE I-4

Main components of dry atmospheric air

Gas	Molecular weight[a]	Molar (or volume) fraction	Mass fraction	Specific gas constant (J kg^{-1} K^{-1})	$m_i R_i/m$ (J kg^{-1} K^{-1})
Nitrogen (N_2)	28.013	0.7809	0.7552	296.80	224.15
Oxygen (O_2)	31.999	0.2095	0.2315	259.83	60.15
Argon (Ar)	39.948	0.0093	0.0128	208.13	2.66
Carbon dioxide (CO_2)	44.010	0.0003	0.0005	188.92	0.09
		1.0000	1.0000		$\dfrac{\Sigma m_i R_i}{m} = 287.05$

[a] Based on 12.000 for C^{12}.

trace components only amounts to less than 0.003%. For all our purposes, which are mainly restricted to the troposphere, we shall consider dry air as a constant mixture which can be treated as a pure gas with a specific constant (see Equation (17) and Table I-4)

$$R_d = 287.05 \text{ J kg}^{-1} \text{ K}^{-1}$$

and a mean molecular weight (see Equation (20)):

$$M_d = R^*/R_d = 0.028964 \text{ kg mol}^{-1} = 28.964 \text{ g mol}^{-1}.$$

The subscript d shall always stand for 'dry air'.

TABLE I-5
Minor gas components of atmospheric air

Gas		Molar (or volume) fraction
Neon	(Ne)	1.8×10^{-5}
Helium	(He)	5.2×10^{-6}
Methane	(CH_4)	1.5×10^{-6}
Krypton	(Kr)	1.1×10^{-6}
Hydrogen	(H_2)	5.0×10^{-7}
Nitrous oxide	(N_2O)	2.5×10^{-7}
Carbon monoxide	(CO)	1.0×10^{-7}
Xenon	(Xe)	8.6×10^{-8}
Ozone	(O_3)	Variable. Up to 10^{-5} in stratosphere.
Sulfur dioxide	(SO_2)	
Hydrogen sulfide and other reduced sulfur compounds	(H_2S, etc.)	Variable, under 10^{-8}
Nitric oxide	(NO)	
Nitrogen dioxide	(NO_2)	
Ammonia	(NH_3)	
Formaldehyde	(CH_2O)	

Amongst the minor components, O_3 has a variable concentration and its molar fraction shows a maximum of 10^{-6} to 10^{-5} with height in the layers around 25 km; it plays an important role in radiation phenomena.

Above about 100 km, the air composition can no longer be considered constant. Firstly, molecular diffusion becomes more important than mixing, which leads to a gradual increase in the proportion of lighter gases with height; secondly, at those heights photochemical reactions produced by solar radiation play an important role in determining the chemical composition. This, however, is of no concern for the subject of this book, which essentially deals with the lower layers of the atmosphere.

Table I-6 gives the value of the second virial coefficient for dry air, and the ratio $pV/R^*T = pv/R_d T$ for two pressures. Virial coefficients of higher order may be neglected, and these data show that we may consider dry air in the troposphere as a perfect gas within less than 0.2% error.

TABLE I-6
Second virial coefficient B' for dry air

$t\,(°C)$	B' (10^{-9} cm^2 dyn^{-1})	$pv/R_d T$	
		$p = 500$ mb	$p = 1000$ mb
-100	-4.0	0.9980	0.9960
-50	-1.56	0.9992	0.9984
0	-0.59	0.9997	0.9994
50	-0.13	0.9999	0.9999

PROBLEMS

1. Suppose we define an empirical scale of temperature t' by Equation (2), based on the saturation vapor pressure e_w (thermometric property) of water (thermometric substance). Construct the graph defining t', using the same 0 and 100 points as for the Celsius temperature t. Starting from this graph, construct a second graph giving t as a function of t' between 0 and 100°.
   What would be the temperature t' corresponding to 50°C?

t(°C)	0	25	50	75	90	100
e_w(mb)	6.11	31.67	123.40	385.56	701.13	1013.25

2. (a) Find the equivalences between MKS and cgs units for the physical quantities listed in Table I-2,
   (b) Show that the definition of the atmosphere is consistent with the original definition in terms of the pressure exerted by a column of 760 mm of Hg at 0°C (density: 13.5951 g cm^{-3}) and standard gravity $g_0 = 9.80665$ m s^{-2}.
3. At what pressure is the ideal gas law in error by 1%, for air at 0°C?
4. Prove that $p = \sum p_i$ and $V = \sum V_i$ (p_i = partial pressure of gas i; V_i = partial volume of gas i) are equivalent statements of Dalton's law.
5. Find the average molecular weight $\bar{M}$ and specific constant $\bar{R}$ for air saturated with water vapor at 0°C and 1 atm of total pressure. The vapor pressure of water at 0°C is 6.11 mb.

CHAPTER II

THE FIRST PRINCIPLE OF THERMODYNAMICS

2.1. Internal Energy

Let us consider a system which undergoes a change while contained in an adiabatic enclosure. This change may be brought about by different processes. For instance, we may increase the temperature T_1 of a given mass of water to a temperature $T_2 > T_1$ by causing some paddles to rotate in the water, or by letting electrical current pass through a wire immersed in the water. In both cases external forces have performed a certain amount of work upon the system.

Experience shows that the work done by external forces adiabatically on the system A_{ad} (that is, the system being enclosed in adiabatic walls) in order to bring about a certain change in its state is independent of the path. In other words, A_{ad} has the same value for every (adiabatic) process causing the same change, and it depends only on the initial and final states of the system. A_{ad} can therefore be expressed by the difference in a state function:

$$A_{ad} = \Delta U \tag{1}$$

and this function U is called the *internal energy* of the system. It follows that for a cycle, $A_{ad} = 0$, and that for an infinitesimal process $\delta A_{ad} = dU$ is an exact differential, which can be expressed by

$$dU = \left(\frac{\partial U}{\partial X}\right)_Y dX + \left(\frac{\partial U}{\partial Y}\right)_X dY \tag{2}$$

as a function of whatever independent variables X, Y are chosen.

The internal energy is defined by (1) except for an arbitrary constant that may be fixed by choosing a reference state. This indetermination is not important, because Thermodynamics only considers the variations in U rather than its absolute value. However, in order to have a unique constant, it is necessary that any state may in principle be related to the same reference state through an adiabatic process. This can always be done. As an example, let us consider an ideal gas. Let E_0 be the chosen reference state, plotted on a p, V diagram (Figure II-1); C is the curve describing the states that may be reached from E_0 by adiabatic expansion or compression. Let us consider any state E_1 of the plane pV at the right of C. E_1 can be reached from E_0 by an infinite number of adiabatic paths; for instance, if the gas is held in an adiabatic container of variable volume (such as an insulating cylinder with a frictionless piston), it can be made to follow the path $E_0 E'$ at $V = $ const., and then $E' E_1$ at $p = $ const.,

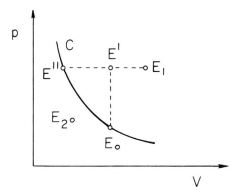

Fig. II-1. Reference state and adiabatic paths.

by having an electric current flow through an inserted resistance while the piston is first held at a fixed position and then left mobile with a constant pressure upon it. Or else it can be made to follow the path $E_0 E''$ by increasing the pressure (with infinite slowness) and the $E'' E_1$ at $p=$const. (as before for the path $E' E_1$).

The points of the plane pV at the left of C, such as E_2, cannot be reached adiabatically from E_0; it can be shown that this would be against the Second Principle. But these points can be related to E_0 be inverting the sense of the process (i.e., reaching E_0 from E_2).

Thus we can fix the arbitrary constant by choosing a single reference state, and relating all other states of the system to that one, for which we set $U=0$ (Figure II-2).

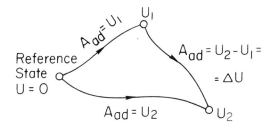

Fig. II-2. Reference state for internal energy.

2.2. Heat

If we now consider a non-adiabatic process causing the same change, we shall find that the work performed on the system will not be the same:

$$A \neq A_{\mathrm{ad}}. \tag{3}$$

We define the *heat* Q absorbed by the system as the difference

$$Q = \Delta U - A. \tag{4}$$

Or, for an infinitesimal process:

$$\delta Q = dU - \delta A. \tag{5}$$

In the adiabatic case, $\delta Q = 0$ by definition.

U is a property of the system, and dU is an exact differential. Neither δA nor δQ are exact differentials, and neither A nor Q are *properties* of the system. In view of the importance of this basic distinction, we shall make at this stage a digression, in order to bring in a short review of related mathematical concepts.

Let δz be a differential expression of the type

$$\delta z = M\,dx + N\,dy \tag{6}$$

where x and y are independent variables, and M and N are coefficients which in general will be functions of x and y. If we want to integrate Equation (6), we shall have the expressions

$$\int M(x, y)\,dx, \quad \int N(x, y)\,dy$$

which are meaningless unless a relation $f(x, y) = 0$ is known. Such a relation prescribes a path in the x, y plane, along which the integration must be performed. This is called a *line integral*, and its result will depend on the given path.

It may be, as a particular case, that

$$M = \frac{\partial z}{\partial x}, \quad N = \frac{\partial z}{\partial y}. \tag{7}$$

We then have

$$\delta z = \frac{\partial z}{\partial x}\,dx + \frac{\partial z}{\partial y}\,dy = dz \tag{8}$$

and δz is therefore an *exact* or *total differential*, which we write dz. In this case, the integration will give

$$\int_1^2 \delta z = z(x_2, y_2) - z(x_1, y_1) = \Delta z \tag{9}$$

or else

$$z = z(x, y) + C \tag{10}$$

where C is an integration constant. z is then a *point function* which depends only on the pair of values (x, y), except for an additive constant. When the integral (9) is taken along a closed curve in the plane x, y, so that it starts and ends at the same point (runs over a cycle), Equation (9) takes the form

$$\oint \delta z = 0. \tag{11}$$

In order to check if a given expression (6) obeys condition (7), it would be necessary to know the function z. It may be easier to apply the theorem of the crossed derivatives

$$\frac{\partial^2 z}{\partial y \, \partial x} = \frac{\partial^2 z}{\partial x \, \partial y} \tag{12}$$

(assuming continuity for z and its first derivatives). That is:

$$\frac{\partial M}{\partial y} = \frac{\partial N}{\partial x}. \tag{13}$$

Equation (13) is therefore a necessary condition for Equation (7) to hold. It is also sufficient for the existence of a function that obeys it, because we can always find the function

$$z = \int M \, dx + \int N \, dy - \iint \frac{\partial M}{\partial y} \, dx \, dy + \text{const.} \tag{14}$$

which, with Equation (13), satisfies Equations (7).

Therefore, Equations (7), (9), (10), (11) and (13) are equivalent conditions that define z as a point function.

If δz is not an exact differential, a factor λ may be found (always in the case of only two independent variables) such that $\lambda \delta z = du$ is an exact differential. λ is called an *integrating factor*.*

The importance that these concepts have for Thermodynamics lies in that state functions like the internal energy are, by their definition, point functions of the state variables. The five conditions above mentioned for z to be a point function find therefore frequent application to thermodynamic functions.

On the other hand, as mentioned before, δA and δQ are not exact differentials.

---

* These concepts may perhaps become clearer through an elementary example. Let

$$\delta z = 2y \, dx + x \, dy$$

which we want to integrate between $x = 0, y = 0$ and $x = 2, y = 2$, and let us choose two arbitrary paths (a) and (b).

(a) The path is defined by $y = x$. Then $\delta z = 3x \, dx$, which is immediately integrable between the two limits, giving $(3/2) x^2 |_0^2 = 6$.

(b) Increase x from 0 to 2 at constant $y = 0$. The integration of δz along this step will give 0. Then increase y to 2 while keeping $x = 2$. This will give 4, which is the total change between the two points, different from the change obtained in (a).

An infinite number of other paths could be devised, with variable results. Obviously, no point function can be defined from δz.

On the other hand, δz admits an integrating factor $\lambda = x$. Thus

$$x \delta z = 2xy \, dx + x^2 \, dy = d(x^2 y),$$

which is an exact differential. Its integral between the limits gives $(2^2 \times 2) - (0 \times 0) = 8$. The same result is obtained if a line integration is performed along the paths (a), (b) or any other. In this case we have a point function, which is $f(x, y) = x^2 y$.

Thus, an expression like

$$A = -\int_i^f p\, dV \tag{15}$$

for the work of expansion, is meaningless if it is not specified how p depends on V along the process.

It is clear, from definition (5) or (4), that exchanging heat is, like performing work, a way of exchanging energy. The definition gives also the procedure for measuring heat in mechanical units. We may cause the same change in a system by an adiabatic process or by a process in which no work is performed. In the first case $\Delta U = A_{ad}$, and in the second, $\Delta U = Q_{A=0}$; as ΔU only depends on the initial and final states, we have $A_{ad} = Q_{A=0}$, which measures $Q_{A=0}$ by the value of A_{ad}. This will be clearer by an example. Let our system be a mass of water, which we bring from 14.5 to 15.5 °C. The first procedure will be to cause some paddles to rotate within the water (as in the classical experiment of Joule) with an appropriate mechanical transmission; A_{ad} may then be measured by the descent of some known weights. The second procedure will be to bring the water into contact with another body at a higher temperature, without performing any work; we say, according to the definition, that $Q_{A=0}$ is the heat gained by our system in this case. If we refer these quantities to one gramme of water, $Q_{A=0}$ is by definition equal to 1 cal, and A_{ad} will turn out to be equal to 4.1855 J. This value is usually known as the *mechanical equivalent of heat*; it amounts to a conversion factor between two different units of energy.

The determination of Q for processes at constant volume or at constant pressure, in which no work is performed upon or by the system except the expansion term (when pressure is kept constant) has been the subject of Calorimetry. Its main experimental results can be briefly summarized as follows:

(1) If in a process at constant pressure no change of physical state and no chemical reactions occur in a homogeneous system, the heat absorbed is proportional to its mass and to the variation in temperature:

$$\delta Q_p = c_p m\, dT \tag{16}$$

where the subindex p indicates that the pressure is kept constant, and the proportionality factor c_p is called the *specific heat capacity* (at constant pressure). The product $c_p m$ is called the *heat capacity* of the system.

Similarly, at constant volume:

$$\delta Q_v = c_v m\, dT. \tag{17}$$

(2) If the effect of the absorption of heat at constant pressure is a change of physical state, which also occurs at constant temperature, Q is proportional to the mass that undergoes the change:

$$\delta Q = l\, dm. \tag{18}$$

The proportionality factor l is called the *latent heat* of the change of state.

(3) If the effect of the absorption of heat, either at constant pressure or at constant volume, is a chemical reaction, Q is proportional to the mass of reactant that has reacted and the proportionality factor is called the *heat of reaction* (referred to the number of moles indicated by the chemical equation) at *constant pressure*, or at *constant volume*, according to the conditions in which the reaction is performed. Heats of reaction have a sign opposite to that of the previous convention, i.e. they are positive if the heat is evolved (lost by the system). We will not be concerned with chemical reactions or their thermal effects.

The specific heat capacity and the latent heat may be referred to the mole instead of to the gramme. They are then called the *molar heat capacities* (or simply *heat capacities*) (at constant pressure C_p and at constant volume C_v) and the *molar heat L* of the change of state considered. In this case, formulas (16)–(18) become

$$\delta Q_p = C_p n \, dT \tag{16'}$$

$$\delta Q_v = C_v n \, dT \tag{17'}$$

$$\delta Q = L \, dn \tag{18'}$$

2.3. The First Principle. Enthalpy

The equation

$$dU = \delta A + \delta Q \tag{19}$$

is the mathematical expression of the First Principle of Thermodynamics.* When applied to an isolated system ($dU=0$), it states the principle of conservation of energy.

We have used the 'egotistical convention' in defining A as the work performed on the system. It is also customary to represent the work done by the system on its surroundings by W. Obviously, $A = -W$.

In the atmosphere we shall only be concerned with one type of work: that of expansion. Therefore

$$\delta A = -p \, dV. \tag{20}$$

It is convenient to define another state function, besides U: the *enthalpy H* (also called *heat content* by some authors)

$$H = U + pV.$$

Introducing these relations, we shall have as equivalent expressions of the first principle:

$$dU = \delta Q - p \, dV \tag{21}$$

$$dH = \delta Q + V \, dp \tag{22}$$

---

* We prefer the more appropriate denomination of '*Principles*' to that commonly used of '*Laws*'.

or their integral expressions

$$\Delta U = Q - \int p\,dV \tag{23}$$

$$\Delta H = Q + \int V\,dp. \tag{24}$$

The same expressions will be used with small letters when referred to the unit mass:

$$du = \delta q - p\,dv \tag{25}$$

etc.

We must remark that Equation (20), and therefore Equations (21) and (22), assume the process to be quasi-static, if p is to stand for the internal pressure of the system. Otherwise, p is the external pressure exerted on the system.

2.4. Expressions of Q. Heat Capacities

Let us consider first a homogeneous system of constant composition. If we now write δQ from Equation (21) or (22), and replace the total differentials dU, dV, dH and dp by their expressions as functions of a chosen pair of variables, we find.

$$\text{Variables } T, V: \quad \delta Q = \left(\frac{\partial U}{\partial T}\right)_V dT + \left[\left(\frac{\partial U}{\partial V}\right)_T + p\right] dV \tag{26}$$

$$\text{Variables } T, p: \quad \delta Q = \left[\left(\frac{\partial U}{\partial T}\right)_p + p\left(\frac{\partial V}{\partial T}\right)_p\right] dT + \left[\left(\frac{\partial U}{\partial p}\right)_T + p\left(\frac{\partial V}{\partial p}\right)_T\right] dp \tag{27}$$

$$\text{Variables } T, p: \quad \delta Q = \left(\frac{\partial H}{\partial T}\right)_p dT + \left[\left(\frac{\partial H}{\partial p}\right)_T - V\right] dp. \tag{28}$$

Equation (27), as well as others that can be derived in a similar way, will be of no particular value to us, but it illustrates, by comparison with Equations (26) and (28), the fact that simpler expressions are obtained when the function U is associated with the independent variable V and when H is associated with p.

For a process of heating at constant volume, we find from Equation (26):

$$C_V = \frac{\delta Q_V}{dT} = \left(\frac{\partial U}{\partial T}\right)_V \tag{29}$$

or

$$c_v = \left(\frac{\partial u}{\partial T}\right)_v. \tag{29'}$$

And for constant pressure, from Equation (28):

$$C_p = \frac{\delta Q_p}{dT} = \left(\frac{\partial H}{\partial T}\right)_p \tag{30}$$

and

$$c_p = \left(\frac{\partial h}{\partial T}\right)_p. \tag{30'}$$

The expressions of δQ also allow us to give values for the heat of change of volume, and of change of pressure, at constant temperature. From Equation (26):

$$\frac{\delta Q_T}{dV} = \left(\frac{\partial U}{\partial V}\right)_T + p$$

and from Equation (28):

$$\frac{\delta Q_T}{dp} = \left(\frac{\partial H}{\partial p}\right)_T - V.$$

2.5. Calculation of Internal Energy and Enthalpy

Equations (29) and (30) can be directly integrated along processes at constant volume and at constant pressure, respectively, to find U and H, if C_v and C_p are known as functions of T:

$$U = \int C_v \, dT + \text{const.} \quad \text{(at constant volume)} \tag{31}$$

$$H = \int C_p \, dT + \text{const.} \quad \text{(at constant pressure)} \tag{32}$$

C_v and C_p, as determined experimentally, are usually given by polynomic expressions, such as

$$C = \alpha + \beta T + \gamma T^2 + \ldots \tag{33}$$

for given ranges of temperatures.

The calculation of U and H for the general case when both T and V (or T and p) change must await consideration of the Second Principle (cf. Chapter III, Section 8).

Let us now consider two rigid containers linked by a connection provided with a stopcock. One of them contains a gas and the other is evacuated. Both are immersed in a common calorimeter. If the stopcock is opened, so that the gas expands to the total volume, it is found that the system (gas contained in both containers) has exchanged no heat with its surroundings. As there is no work performed (the total volume of the system remained constant), we have*:

$$Q = 0; \quad A = 0; \quad \Delta U = 0.$$

Actually this experiment, which was performed by Joule, was later improved by Joule and Thomson, who found a small heat exchange (Joule-Thomson effect). However, the effect vanishes for ideal gas behavior. In fact, if the equation of state $pV = R^*T$ is accepted as the definition of ideal gas, it may be shown, with the aid of the second principle, that this must be so**. Therefore the previous result is exact for ideal gases. As p changed during the process, we conclude that U for an ideal gas is only a function of T:

$$U = U(T)$$

and the partial derivatives of the previous formulas become total derivatives:

(34) $\quad C_V = \dfrac{dU}{dT}; \quad c_v = \dfrac{du}{dT}.$ \hfill (34′)

Similarly, $H = U + pV = U + R^*T = H(T)$; therefore

(35) $\quad C_p = \dfrac{dH}{dT}; \quad c_p = \dfrac{dh}{dT}.$ \hfill (35′)

---

* As the gas in the container, where it was originally confined, expands into the other, work is done by some portions of the gas against others, while their volumes are changing. These are internal transfers not to be included in A. This is an example demonstrating that systems must be defined carefully and clearly when considering a thermodynamic process; the system in this case is best defined as all the gas contained within the *two* containers (whose total volume is constant).

** It will be seen (Chapter III, Section 5) that for a reversible process

$$dU = T\,dS - p\,dV = T\,dS - R^*T\dfrac{dV}{V}$$

where S is the entropy. Dividing by T:

$$\dfrac{dU}{T} = dS - R^*\,d\ln V.$$

dS is an exact differential, by the Second Law, and so is the last term. Therefore (dU/T) is also an exact differential, which may be written (by developing dU):

$$\dfrac{dU}{T} = \dfrac{1}{T}\left(\dfrac{\partial U}{\partial T}\right)_p dT + \dfrac{1}{T}\left(\dfrac{\partial U}{\partial p}\right)_T dp.$$

Applying to (dU/T) the condition of equality of crossed second derivatives

$$\dfrac{1}{T}\dfrac{\partial^2 U}{\partial p\,\partial T} = -\dfrac{1}{T^2}\left(\dfrac{\partial U}{\partial p}\right)_T + \dfrac{1}{T}\dfrac{\partial^2 U}{\partial T\,\partial p}$$

or

$$\left(\dfrac{\partial U}{\partial p}\right)_T = 0$$

and

$$U = U(T).$$

It can readily be derived that

(36) $\quad C_p - C_v = R^*; \qquad c_p - c_v = R.$ \hfill (36')

Heat capacities of gases can be measured directly, and the corresponding coefficients for introduction into Equation (33) can be determined experimentally in order to represent the data over certain temperature intervals. For simple gases as N_2, O_2, Ar, however, the data are nearly constant for all the ranges of temperature and pressure values in which we are interested. This agrees with the theoretical conclusions from Statistical Mechanics, which indicate the following values (*):

$$\text{Monatomic gas:} \quad C_v = \tfrac{3}{2}R^*; \quad C_p = C_v + R^* = \tfrac{5}{2}R^*$$

$$\text{Diatomic gas:} \quad C_v = \tfrac{5}{2}R^*; \quad C_p = \tfrac{7}{2}R^*.$$

We shall define the ratio coefficients

$$\varkappa = R^*/C_p = R/c_p; \qquad \eta = C_p/C_v = c_p/c_v.$$

For dry air, considered as a diatomic gas (neglecting the small proportion of Ar, CO_2, and minor components), we should expect:

$$\varkappa_d = 2/7 = 0.286; \qquad \eta_d = 7/5 = 1.400$$

$$c_{vd} = 718 \text{ J kg}^{-1}\text{K}^{-1} = 171 \text{ cal kg}^{-1}\text{K}^{-1}$$

$$c_{pd} = 1005 \text{ J kg}^{-1}\text{K}^{-1} = 240 \text{ cal kg}^{-1}\text{K}^{-1}.$$

These values are in good agreement with experience, as can be seen from the values of c_{pd} from Table II-1.

---

* According to the principle of equipartion of energy, the average molecular energy is given by

$$\bar{\varepsilon} = rkT/2$$

where r is the number of squared terms necessary to express the energy:

$$\varepsilon = \sum_1^r \tfrac{1}{2}\lambda_i \xi_i^2$$

k is Boltzmann's constant, λ_i are constants, and ξ_i are generalized coordinates or momenta. If no potential energy has to be considered, the ξ_i are all momenta and the number r is equal to the *degrees of freedom* of the molecules, as the number of generalized coordinates needed to determine their position: 3 for a monatomic gas, and two more (angular coordinates to give the orientation) for diatomic molecules; it is here assumed that the effect of vibration of the diatomic molecules may be neglected, which is true for N_2 and O_2 in the range of temperatures in which we are interested.

$$U = N_A \bar{\varepsilon} = \frac{r}{2} R^* T$$

$$C_v = \frac{dU}{dT} = \frac{r}{2} R^*$$

(N_A = Avogadro's number).

TABLE II-1

c_{p_d} in IT cal kg^{-1} K^{-1}

p(mb)	t(°C)			
	-80	-40	0	$+40$
0	239.4	239.5	239.8	240.2
300	239.9	239.8	239.9	240.3
700	240.4	240.1	240.1	240.4
1000	241.0	240.4	240.3	240.6

For simple ideal gases in general, and for a wide range of temperatures, the specific heats can be considered as constant, and the expressions of the internal energy and the enthalpy can be obtained by integrating Equations (34) and (35):

$$U = C_v T + a \tag{37}$$

$$H = C_p T + a. \tag{38}$$

Taking $H - U$, and noticing that

$$(C_p - C_v)T = R^* T = pV,$$

it may be seen that the additive constant a, although arbitrary, must be the same for both functions.

Within this approximation, the two expressions (21) and (22) of the First Principle may be written as

$$\delta Q = C_v \, dT + p \, dV \tag{39}$$

and

$$\delta Q = C_p \, dT - V \, dp. \tag{40}$$

2.6. Latent Heats of Pure Substances. Kirchhoff's Equation

In general, from the expressions (21) and (22) of the First Principle, we can see that the heat absorbed by a system in a reversible process at constant volume Q_v (the subscript indicating the constancy of that variable) is measured by the change in internal energy and the heat at constant pressure Q_p by the change in enthalpy:

$$\delta Q_v = dU \tag{41}$$

$$\delta Q_p = dH. \tag{42}$$

In the case of homogeneous systems, these formulas would give $dU = nC_v \, dT = mc_v \, dT$ and $dH = nC_p \, dT = mc_p \, dT$ (cf. Equations (39) and (40)). However, we are now interested in changes of phase. As we shall see later, only one independent variable is left for a system made out of a pure substance, when two phases are in equilibrium. Thus if we fix the pressure or the volume, the temperature at which the change of physical state may occur reversibly is also fixed, and the specific heats need not be considered in these processes.

The latent heats are defined for changes at constant pressure. Therefore, in general

$$L = \Delta H \tag{43}$$

or

$$l = \Delta h. \tag{43'}$$

In particular L may be L_f, L_v, L_s = molar heats of fusion, vaporization, sublimation. Similarly l can be any of the latent heats l_f, l_v, l_s.

It is of interest to know how L varies with temperature. We may write

$$dH = \left(\frac{\partial H}{\partial T}\right)_p dT + \left(\frac{\partial H}{\partial p}\right)_T dp \tag{44}$$

for two states, a and b, such that $\Delta H = L = H_b - H_a$. Taking the difference of both exact differentials, we have

$$d(\Delta H) = \left(\frac{\partial \Delta H}{\partial T}\right)_p dT + \left(\frac{\partial \Delta H}{\partial p}\right)_T dp. \tag{45}$$

If we now assume that p is maintained constant, only the first term on the right is left, and we have

$$d(\Delta H)_p = dL = \left(\frac{\partial \Delta H}{\partial T}\right)_p dT = \left(\frac{\partial H_b}{\partial T}\right)_p dT - \left(\frac{\partial H_a}{\partial T}\right)_p dT = (C_{p_b} - C_{p_a}) dT$$

or

$$(\partial L/\partial T)_p = \Delta C_p. \tag{46}$$

This is called Kirchhoff's equation. If the heat capacities are known as empirical functions of the temperature expressed as in Equation (33), L can be integrated in the same range as

$$L = \int \Delta C_p \, dT$$

$$= L_0 + \Delta \alpha T + \frac{\Delta \beta}{2} T^2 + \frac{\Delta \gamma}{3} T^3 + \ldots \tag{47}$$

where L_0 is an integration constant, and the Δ are always taken as differences between phases a and b.

The physical sense of Kirchhoff's equation may perhaps become clearer by consideration of the cycle indicated below. By equating the change in enthalpy from a to b at the temperature T for the two paths indicated by the arrows, Equation (46) is again obtained.

It may be remarked that Kirchhoff's equation holds true also for reaction heats in thermochemistry, and that another similar equation may be derived for the change

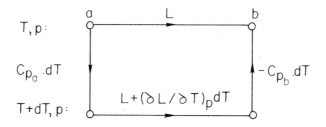

in heat of reaction or of change of physical state with temperature at constant volume, where the internal energy U and the heat capacities at constant volume C_v play the same role as H and C_p in the above derivation:

$$\left(\frac{\partial \Delta U}{\partial T}\right)_v = \Delta C_v. \tag{48}$$

It should be remarked that the variation of L with T alone, i.e., while keeping p constant, cannot correspond to changes in which equilibrium conditions between the two phases are maintained. This would require a simultaneous variation of pressure, related to the temperature by the curve of equilibrium for the change of phase. To treat this problem requires the help of the Second Principle; this will be done in Chapter IV, Section 8, where the last term on the right in Equation (44) will be calculated. For vaporization and sublimation, however, it will be shown that this term is entirely negligible, while for fusion it should be taken into account, according to the formula

$$\frac{dL_f}{dT} = \Delta C_p + \frac{L_f}{T}\left[1 - \frac{T}{\Delta V}\left(\frac{\partial \Delta V}{\partial T}\right)_p\right] \tag{49}$$

where $\Delta V = V_{\text{liquid}} - V_{\text{solid}}$.

2.7. Adiabatic Processes in Ideal Gases. Potential Temperature

Considering Equations (39) and (40) we may write, for an adiabatic process in ideal gases:

$$\delta Q = 0 = C_v \, dT + p \, dV =$$
$$= C_p \, dT - V \, dp. \tag{50}$$

Dividing by T and introducing the gas law we derive the two first following equations:

$$0 = C_v \, d \ln T + R^* \, d \ln V$$
$$= C_p \, d \ln T - R^* \, d \ln p$$
$$= C_v \, d \ln p + C_p \, d \ln V. \tag{51}$$

The third equation results from any of the other two by taking into account that $d \ln p + d \ln V = d \ln T$ (by taking logarithms and differentiating the gas law) and

Equation (36). Integration of the three equations gives

$$T^{C_v}V^{R^*} = \text{const.}$$
$$T^{C_p}p^{-R^*} = \text{const.} \tag{52}$$
$$p^{C_v}V^{C_p} = \text{const.}$$

Or, introducing the ratios $\varkappa$ and η:

$$TV^{\eta-1} = \text{const.}$$
$$Tp^{-\varkappa} = \text{const.} \tag{53}$$
$$pV^{\eta} = \text{const.}$$

These are called Poisson's equations. They are equivalent, being related one to the other by the gas law. The third equation may be compared with Boyle's law for isotherms: $pV = \text{const}$. In the p, V plane the *adiabats* (curves representing an adiabatic process) have larger slopes than the isotherms, due to the fact that $\eta > 1$. This is shown schematically in Figure II-3b. Figure II-3a shows the isotherms and adiabats on the three-dimensional surface representing the states of an ideal gas with coordinates p, V, T. Projections of the adiabats on the p, T and V, T planes are given in Figure II-3c and d, which also indicate the isochores and isobars.

If we apply the second Equation (53) between two states, we have:

$$\frac{T_0}{T} = \left(\frac{p_0}{p}\right)^{\varkappa}. \tag{54}$$

If we choose p_0 to be 1000 mb, T_0 becomes, by definition, the *potential temperature* θ:

$$\theta = T\left(\frac{1000 \text{ mb}}{p}\right)^{\varkappa}. \tag{55}$$

The potential temperature of a gas is therefore the temperature that it would take if we compressed or expanded it adiabatically to the pressure of 1000 mb. We shall see that this parameter plays an important role in Meteorology.

The importance of potential temperature in meteorological studies is directly related to the fundamental role of adiabatic processes in the atmosphere. If we restrict our attention to dry air, we may assert that only radiative processes cause addition to or abstraction of heat from a system consisting of a unique sample of the atmosphere. In general, however, we must deal with bulk properties of the atmosphere, i.e., averaged properties, and in such an open system we recognize that three-dimensional mixing processes take place into and out of any system moving with the bulk flow. To this extent, then, we must add turbulent diffusion of heat to our non-adiabatic processes. Nevertheless, except in the lowest 100 mb of the atmosphere, these non-adiabatic processes are relatively unimportant and it is generally possible to treat changes of state as adiabatic, or at least as quasi-adiabatic.

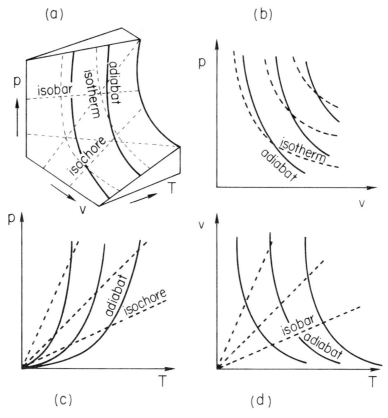

Fig. II-3. Thermodynamic surface for ideal gases and projections on p, v; p, T and v, T planes.

As can be seen from Equations (54) and (55), θ is a constant for an insulated gaseous system of fixed composition, i.e., for an adiabatic process. This constant may be used to help specify a particular system, such as a volume element of dry air. Its value does not change for any adiabatic process, and we say that potential temperature is conserved for adiabatic processes. Conservative processes are important in Meteorology since they enable us to trace the origin and subsequent history of air masses and air parcels, acting as tags (or tracers). If air moves along an isobaric surface (p constant), the temperature of the air sample will not change, if no external heat source exists. In general, air motion is very nearly along isobaric surfaces, but the small component through isobaric surfaces is of great importance. If the pressure of an air sample changes, then its temperature will change also, to maintain a constant value for the potential temperature. Changes of pressure and temperature will have the same signs; thus adiabatic compression is accompanied by warming and adiabatic expansion by cooling. Adiabatic compression, with pressure increasing along a trajectory, usually implies that the air is sinking or subsiding or descending (all these terms are employed in Meteorology), whereas adiabatic expansion, with the pressure on an air sample or element decreasing with time, usually implies ascent.

During adiabatic ascent, as the temperature falls the relative humidity rises (if the air contains any water vapor). Eventually a state of saturation is reached and further ascent causes condensation, releasing the latent heat of condensation which tends to warm the air and to change its potential temperature. Strictly speaking, this is still an adiabatic process, since the heat source is internal rather than external to the system. However, it is clear that our formulation of the first law for gases is no longer valid if phase changes occur. It is for this reason that the potential temperature is not conservative for processes of evaporation or condensation, regardless of whether the heat source (for the latent heat) is internal or external (for evaporation from a ground or water source, for example). We shall see later that the presence of unsaturated water vapor has no significant effect on the conservation of the (dry air) potential temperature.

For a non-adiabatic process, it is possible to evaluate the change in potential temperature to be expected, since by definition the potential temperature cannot be conserved during non-adiabatic processes. Let us take natural logarithms of Equation (55), and then differentiate:

$$d \ln \theta = d \ln T - \varkappa \, d \ln p. \tag{56}$$

Using Equations (51) and (55), we obtain

$$\delta Q = C_p T \, d \ln \theta = C_p \left(\frac{p}{p_0} \right)^{\varkappa} d\theta. \tag{57}$$

These are, of course, merely additional formulations of the First Principle, for ideal gases.

2.8. Polytropic Processes

Although vertical motions can be generally considered to be adiabatic, there is a special type of situation where this approximation ceases to be valid. This is when very slow motions of horizontally-extended atmospheric layers are associated with some exchange of energy by radiation; the problem will be considered in Chapter VII, Section 11. Here we point out that such a process can be approximated by writing

$$\delta Q = C \, dT \quad \text{or} \quad \delta q = c \, dT \tag{58}$$

with a constant C or c, which can be called the polytropic molar or the polytropic specific heat capacity, respectively, associated with the process. By definition this is called a *polytropic process*.

Instead of (5), we must now write

$$C_v \, dT + p \, dV = C_p \, dT - V \, dp = C \, dT \tag{59}$$

or

$$(C_v - C) \, dT + p \, dV = (C_p - C) \, dT - V \, dp = 0. \tag{60}$$

Repeating the derivation that led to (53), we find now:

$$TV^{n-1} = \text{const.}$$
$$Tp^{-k} = \text{const.} \tag{61}$$
$$pV^n = \text{const.}$$

where the 'polytropic exponent'

$$n = \frac{C_p - C}{C_v - C} = \frac{c_p - c}{c_v - c}$$

replaces η in the previous formulas and

$$k = \frac{n-1}{n} = \frac{R^*}{C_p - C} = \frac{R}{c_p - c}.$$

replaces κ in the previous formulas.

Any of the parameters c (or C), n and k can be used to characterize the polytropic process. We can easily see that, as particular cases, $C = C_p$ or $n = 0$ or $k = -\infty$ corresponds to an isobaric process; $C = \infty$ or $n = 1$ or $k = 0$ corresponds to an isothermal process; $C = 0$ or $n = \eta$ or $k = \varkappa$ corresponds to an adiabatic process; and $C = C_v$ or $n = \infty$ or $k = 1$ corresponds to an isochoric process.

PROBLEMS

1. Calculate A and Q for an isothermal compression (isothermal change, which may or may not be brought about by an isothermal process) of an ideal gas from A_i to A_f (see Figure) for each of the four following processes:

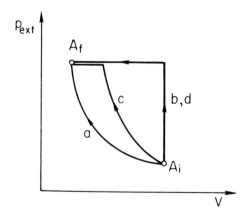

(a) isothermal reversible compression.
(b) sudden compression to $p_{ext}=p_f$ (e.g. dropping a weight on the piston of a cylinder containing the gas) and the subsequent contraction.
(c) adiabatic reversible compression to p_f followed by reversible isobaric cooling.
(d) reversible increase of the temperature at constant volume until $p=p_f$, followed by reversible decrease of temperature at constant pressure until $V=V_f$.

p_{ext} denotes the external pressure applied to the system.

2. A dry air mass ascends in the atmosphere from the 1000 mb level to that of 700 mb. Assuming that it does not mix and does not exchange heat with its surroundings, and that the initial temperature is 10 °C, calculate:
   (a) Its initial specific volume.
   (b) Its final temperature and specific volume.
   (c) Its change in specific internal energy and in specific enthalpy (in J kg^{-1} and in cal g^{-1}).
   (d) What is the work of expansion done by 1 km^3 of that air (volume taken at initial pressure)?
   (e) What would the specific enthalpy change have been, for an isobaric cooling to the same final temperature, and for an isothermal expansion to the same final pressure?
   (f) Compute (a), (b) and (c) for pure Ar, instead of dry air. (Atomic weight of Ar: 39.95.)

3. The figure represents an insulated box with two compartments A and B, each containing a monatomic ideal gas. They are separated by an insulating and perfectly flexible wall, so that the pressure is equal on both sides. Initially each compartment measures one liter and the gas is at 1 atm and 0 °C. Heat is then supplied to gas A (e.g. by means of an electrical resistance) until the pressure rises to 10 atm. Calculate:
   (a) The final temperature T_B.
   (b) The work performed on gas B.
   (c) The final temperature T_A.
   (d) The heat Q_A absorbed by gas A.

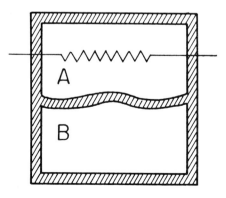

4. Consider one mole of air at 0°C and 1000 mb. Through a polytropic process, it acquires three times its initial volume at 250 mb. Calculate:
   (a) The value of n in $pV^n = $ const.
   (b) The final temperature.
   (c) The change in internal energy.
   (d) The work received by the gas.
   (e) The heat absorbed by the gas.
   Consider that the air behaves as an ideal gas.

5. One gram of water is heated from 0 to 20°C, and then evaporated at constant temperature (at the vapour pressure of water corresponding to that temperature). Compute, in cal g^{-1} and cal g^{-1} K^{-1},
   (a) Δu
   (b) Δh
   (c) The mean value of c_{p_v} between 0 and 20°C, knowing that

   $$l_{v,0°C} = 597.3 \text{ cal g}^{-1}, \quad l_{v,20°C} = 586.0 \text{ cal g}^{-1}, \quad c_w = 1.00 \text{ cal g}^{-1} \text{K}^{-1}$$

   (c_{p_v}: specific heat capacity of water vapor at constant pressure; c_w: specific heat capacity of liquid water).

   Note–Assume that any variation in pressure has a negligible effect and that water vapor behaves as an ideal gas.

CHAPTER III

THE SECOND PRINCIPLE OF THERMODYNAMICS

3.1. The Entropy

The first principle of thermodynamics states an energy relation for every process that we may consider. But it does not say anything about whether this process might actually occur at all. It is the second principle that faces this problem and provides at the same time a rigorous criterion to decide when a system is in a state of thermodynamic equilibrium.

Thermodynamic processes may be classified into three categories: natural, impossible or antinatural and reversible.

Natural processes are always in greater or lesser degree, irreversible, and the direction in which they occur is obviously towards equilibrium, by definition of equilibrium. Expansion of a gas into a vacuum, heat conduction through a finite temperature gradient, combination of oxygen and hydrogen at room temperature producing water, diffusion of one gas into another under a finite concentration gradient, freezing of supercooled water – these are examples of natural, irreversible processes. We might imagine the reverse processes: contraction of a gas under no external pressure, heat flowing from one body to another at a higher temperature, etc. But these, as well as other more complicated processes whose absurdity would seem less obvious, are impossible in nature; their impossibility is prescribed by the second principle.

Natural processes may produce opposite changes in a system, depending on the external conditions. We can, in general, reduce the irreversibility of these processes by modifying their paths so that the difference between the actual values of the state variables and the values that would correspond to an equilibrium state is reduced through all the stages of the process. If we continue doing this indefinitely, we tend to a common limit for processes producing either one change or its opposite. This limit is called a reversible process, as we have already seen in Chapter I, Section 8. A reversible process is thus an ideal limit, which cannot actually be realized but to which one can approximate indefinitely, and may be defined as a series of states that differ infinitesimally from equilibrium and succeed each other infinitely slowly, while the variables change in a continuous way. As it may be noticed, we have assumed that, according to experience, the rate of the process tends to zero as the conditions tend to that of equilibrium.

Referring to the previous examples, expansion of a gas with pressure p against an external pressure $p - dp$, heat conduction along an infinitesimal temperature gradient, etc., are reversible processes. By reversing the sign of the infinitesimal difference from equilibrium conditions (replacing dp by $-dp$, dT/dx by $-dT/dx$, etc.), the reversible process of the opposite sense will occur.

We make now the following statement: for every reversible process there is a positive integrating* factor $1/\tau$ of the differential expression δQ, which is only dependent on the thermal state and which is equal for all thermodynamic systems. A state function S is thus defined (cf. Chapter I, Section 2), the *entropy*, by the exact differential dS:

$$dS = \left(\frac{\delta Q}{\tau}\right)_{\text{rev}}. \tag{1}$$

We further state that for irreversible processes

$$dS > \left(\frac{\delta Q}{\tau}\right)_{\text{irr}}. \tag{2}$$

So that for any process we may write

$$dS \geq \left(\frac{\delta Q}{\tau}\right) \tag{3}$$

and the inequality and equality symbols shall always correspond to irreversible and reversible processes, respectively. This formula is the mathematical expression of the second principle of thermodynamics.

It may be shown that τ turns out to be proportional to the absolute ideal gas temperature in all its range of validity, and as the proportionality factor is arbitrary, the simplest choice is to take $\tau = T$. We shall now consider this problem.

3.2. Thermodynamic Scale of Absolute Temperature

Let us first consider a Carnot cycle performed by an ideal gas. The cycle is reversible and consists of two isotherms, at temperatures T_1 and $T_2 < T_1$, and two adiabats (Figure III-1).

Let us compute the work A and the heat Q for the four steps (in the sense indicated by the arrows):

(I) $\quad \Delta U_{\text{I}} = 0; \quad Q_1 = -A_{\text{I}} = \int_A^B p\,dV = nR^*T_1 \ln\frac{V_B}{V_A}$

(II) $\quad Q_{\text{II}} = 0; \quad -A_{\text{II}} = -\Delta U_{\text{II}} = C_v(T_1 - T_2)$

(III) $\quad \Delta U_{\text{III}} = 0; \quad Q_2 = -A_{\text{III}} = -nR^*T_2 \ln\frac{V_C}{V_D}$

(IV) $\quad Q_{\text{IV}} = 0; \quad -A_{\text{IV}} = -\Delta U_{\text{IV}} = -C_v(T_1 - T_2)$.

---

* It might have been taken negative. This is a matter of convention.

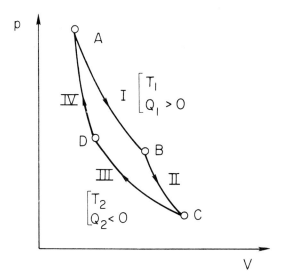

Fig. III-1. Carnot's cycle.

The sum of the four terms ΔU is zero, as it should be for a cycle. The work terms in the two adiabats cancel each other. The gas absorbs the quantity of heat $Q_1 > 0$ from the hotter reservoir and rejects $Q_2 < 0$ to the colder one. Furthermore, from Poisson's relations

$$\frac{T_1}{T_2} = \left(\frac{V_C}{V_B}\right)^{\eta-1} = \left(\frac{V_D}{V_A}\right)^{\eta-1}$$

we derive

$$V_B/V_A = V_C/V_D \tag{4}$$

which, introduced in the expressions for Q_1 and Q_2, gives:

$$\frac{Q_1}{T_1} + \frac{Q_2}{T_2} = 0 \tag{4}$$

or

$$\frac{|Q_2|}{Q_1} = \frac{T_2}{T_1}. \tag{5}$$

The application of Equation (1) to the same cycle would give, considering that τ must remain constant in each isotherm:

$$\frac{Q_1}{\tau_1} + \frac{Q_2}{\tau_2} = 0 \tag{6}$$

or

$$\frac{|Q_2|}{Q_1} = \frac{\tau_2}{\tau_1}. \tag{7}$$

Therefore, if we write

$$\frac{\tau_2}{\tau_1} = \frac{T_2}{T_1}, \tag{8}$$

Equation (1) is obeyed.

This shows that τ must be taken as proportional to T (absolute ideal gas temperature as defined in Chapter I, Section 5) for the ideal gas of the Carnot cycle.

In order to generalize this result, we should follow the usual procedure by which the mathematical expression of the second principle is shown to derive from physical considerations based on the impossibility of a perpetuum mobile of the second kind (cf. Section 4). As a first step, we should show that Equation (4) is also valid for any reversible cycle performed between two heat sources T_1 and T_2, independently of the nature of the cycle and of the systems (Carnot's theorem), while for an irreversible cycle, $(Q_1/T_1) + (Q_2/T_2) < 0$. Finally we might show that any other cycle may be decomposed into, or referred to, a number of Carnot cycles (reversible cycles between two isotherms and two adiabats performed by any fluid), which is made to tend to infinity if the temperature varies continuously for the main cycle. We shall omit here these derivations, for which we refer to any good textbook on general thermodynamics. We shall only quote the final result, which is the formula

$$\oint \frac{\delta Q}{T} \leqslant 0 \tag{9}$$

where the circle in the integral sign indicates that it refers to a cycle and the inequality sign corresponds, as always, to the irreversible case. And, as from Equation (3) it follows that in every case

$$\oint \frac{\delta Q}{\tau} \leqslant 0 \tag{10}$$

we conclude that τ is always proportional, and may be put equal, to T:

$$\tau = T. \tag{11}$$

The equality is achieved by choosing again the value 273.16 K for the reference state, namely the triple point of water. We have now the thermodynamic or Kelvin scale of temperature defined by

$$T = 273.16 \frac{|Q|}{|Q_t|} \tag{12}$$

where Q_t is the heat taken from, or given to, a source in thermal equilibrium with the triple point of water, in a reversible cycle between this source and any other one, to which correspond the values Q and T. As the cycle can be performed, in principle, by any system, this definition of temperature is completely general, covering any possible thermal states, and coincides with the ideal gas absolute temperature for all the range covered by this one.

3.3. Formulations of the Second Principle

We shall, therefore, write the second principle as

$$dS \geqslant \frac{\delta Q}{T}. \tag{13}$$

The entropy is thus defined (through Equation (13) with the equality sign, for reversible processes) as a state function and, just as in the case of the internal energy, it has an arbitrary additive constant. This constant can be fixed by choosing a reference state for which we write $S=0$. The third principle of thermodynamics provides an absolute reference state (temperature of zero Kelvin for all perfect crystalline solids), but its study, essential in chemical thermodynamics, can be omitted in this book of applications to the atmosphere.

It should be remarked that, as was the case for pressure, the temperature of the system is not defined for irreversible processes. The meaning of T in this case is the temperature of the heat sources in contact with the system.

Other expressions of the second principle for particular cases are:

finite process: $\qquad \Delta S \geqslant \int \frac{\delta Q}{T}$ \hfill (14)

adiabatic process: $\qquad dS \geqslant 0$ \hfill (15)

isentropic process: $\qquad 0 \geqslant \delta Q$ \hfill (16)

finite isothermal process: $\quad \Delta S \geqslant \frac{Q}{T}$ \hfill (17)

cycle: $\qquad \oint \frac{\delta Q}{T} \leqslant 0.$ \hfill (18)

Their derivation from Equation (13) is trivial. Equation (15) tells us that if a process is adiabatic and reversible, it is also isentropic.

If in any process we consider a larger system including the primitive system plus all the bodies with which it exchanged heat, the process becomes adiabatic for the total system and, according to Equation (15), $dS_{\text{total}} \geqslant 0$. In that sense it is sometimes said, rather loosely, that the entropy 'of the universe' increases in every natural process.

3.4. Lord Kelvin's and Clausius' Statements of the Second Principle

If we consider a system performing a cycle in such a way that at the end it has exchanged heat with only one source T, Equation (18) gives

$$\frac{Q}{T} \leqslant 0$$

and, as $T>0$, $Q \leqslant 0$. On the other hand, from the first principle it follows that $A = -Q$.

It is, therefore, impossible to construct a cyclic device that will produce work ($A<0$) and no other effect than the cooling of a unique heat source ($Q>0$). This impossibility of a "*perpetuum mobile* of the second kind" is Lord Kelvin's statement of the second principle.

If we now consider a system performing a cycle between two sources T_1 and $T_2 < T_1$, without any production or consumption of work, we shall have

$$\frac{Q_1}{T_1} + \frac{Q_2}{T_2} \leqslant 0$$

but by hypothesis and the first law ($A=0$, $Q_1+Q_2=0$), $Q_1 = -Q_2 = Q$. It follows that $Q = Q_1 \geqslant 0$ (and $Q_2 \leqslant 0$). Therefore, it is impossible to construct a cyclic device that will produce the sole effect of transferring heat from a colder to a hotter reservoir. This is Clausius' statement of the second principle.

Therefore these two traditional statements, which are alternative bases for the development of the second principle and its applications, appear as immediate consequences of the mathematical formulation in Equation (13).

3.5. Joint Mathematical Expressions of the First and Second Principles. Thermodynamic Potentials

By introducing Equation (13) in the expressions of the first principle (Chapter II, Section 3) we obtain

$$dU \leqslant T\,dS + \delta A \tag{19}$$

and for the case in which $\delta A = -p\,dV$:

$$dU \leqslant T\,dS - p\,dV \tag{20}$$

$$dH \leqslant T\,dS + V\,dp. \tag{21}$$

In the last two expressions, U and H appear as functions of S and V, and S and p, respectively, as independent variables. It is also convenient to have general expressions where the pairs T, V and T, p appear as the independent variables. Two new state functions are defined with that purpose:

$$\text{Helmholtz function or free energy}^* = F = U - TS \tag{22}$$

$$\text{Gibbs function or free enthalpy}^* = G = H - TS$$

$$= U + pV - TS. \tag{23}$$

Both functions are also called thermodynamic potentials. By differentiating F

---

* These are the denominations and symbols recommended internationally. Helmhotz function has also been called *work function* and represented by the letter A. Gibbs function has also been known as the *free energy* and represented by F. Thus Helmholtz function is sometimes referred to as *Helmholtz free energy*, in order to avoid ambiguity.

and introducing it in Equation (20), and by differentiating G and introducing it in Equation (21), two new joint expressions of both thermodynamic principles are obtained:

$$dF \leqslant -S\,dT - p\,dV \tag{24}$$

$$dG \leqslant -S\,dT + V\,dp. \tag{25}$$

We shall make use mainly of the last one. In all four Equations (20), (21), (24), (25), the inequality sign holds for irreversible processes (p is then the external pressure exerted upon the system, and T the temperature of the heat sources in contact with it) and the equality sign for reversible processes.

We may summarize these expressions or *fundamental equations* in the following table, where U, H, F and G are considered as the *characteristic functions*:

Characteristic function	Independent variables	Fundamental equation
U	S, V	$dU \leqslant T\,dS - p\,dV$
H	S, p	$dH \leqslant T\,dS + V\,dp$
F	T, V	$dF \leqslant -S\,dT - p\,dV$
G	T, p	$dG \leqslant -S\,dT + V\,dp$

(26)

We see that these expressions have a helpful symmetry, where the choice of independent variables can be represented schematically by

$$
\begin{array}{ccc}
U & S & H \\
V & & p \\
F & T & G.
\end{array}
$$

Each characteristic function is associated with the two adjacent variables. The convenience of these particular associations becomes apparent through the simplicity of many derivations and the equilibrium conditions. It should be noticed, however, that this symmetry only holds under the restriction $\delta A = -p\,dV$; thus, if this restriction was not imposed, we should substitute Equations (21) and (25) by

$$dH \leqslant T\,dS + V\,dp + p\,dV + \delta A \tag{27}$$

$$dG \leqslant -S\,dT + V\,dp + p\,dV + \delta A \tag{28}$$

and the advantages of defining H and G would be lost.

All these thermodynamic functions (U, H, S, F, G) are state functions to which the mathematical concepts reviewed in Chapter II, Section 2 apply. In particular, we can apply Equations (1) and (13) from Chapter II to the differential expressions in

Equations (26). The first ones give:

$$\left(\frac{\partial U}{\partial S}\right)_V = \left(\frac{\partial H}{\partial S}\right)_p = T$$

$$\left(\frac{\partial F}{\partial T}\right)_V = \left(\frac{\partial G}{\partial T}\right)_p = -S$$

$$\left(\frac{\partial U}{\partial V}\right)_S = \left(\frac{\partial F}{\partial V}\right)_T = -p \tag{29}$$

$$\left(\frac{\partial H}{\partial p}\right)_S = \left(\frac{\partial G}{\partial p}\right)_T = V.$$

And applying (Chapter II – Equation (13)), the so-called Maxwell relations are obtained:

$$\left(\frac{\partial T}{\partial V}\right)_S = -\left(\frac{\partial p}{\partial S}\right)_V; \qquad \left(\frac{\partial S}{\partial V}\right)_T = \left(\frac{\partial p}{\partial T}\right)_V$$

$$\left(\frac{\partial T}{\partial p}\right)_S = \left(\frac{\partial V}{\partial S}\right)_p; \qquad \left(\frac{\partial S}{\partial p}\right)_T = -\left(\frac{\partial V}{\partial T}\right)_p. \tag{30}$$

Other relations could also be obtained by combining Equations (29) and (30).

We could write the mathematical expression of the second principle in Equation (13) in the alternative way

$$dS = \frac{\delta Q}{T} + \frac{\delta Q'}{T} \tag{31}$$

with

$$\delta Q' \geq 0. \tag{32}$$

With this definition, the inequality sign in the fundamental Equations (26) could be substituted by the subtraction of $\delta Q'$ from the right hand side; for instance, the first equation would read

$$dU = T\,dS - p\,dV - \delta Q'. \tag{33}$$

$\delta Q'$ is called *Clausius' non-compensated heat*, and gives a measure of the irreversibility of the process.

It should be understood that the set (29) are always true, whether we are considering reversible or irreversible processes (both derivatives and right hand sides are state functions or state variables depending only on the state of the system and not on the path under consideration), but T and p are defined only in a system in conditions of equilibrium. Thus in (26), when the process is irreversible and the inequality sign prevails, T and p represent the temperature of the sources in contact with the system and the external pressure on the system. The conditions of irreversibility are established by the differences between these and the equilibrium values: $(T - T_{eq})$ (thermal irreversibility), $(p - p_{eq})$ (mechanical irreversibility). If we equate dU given by (33) for an irreversible path with the first (26) when applied to the same change by a reversible path, we find

$$\delta Q' = (T - T_{eq})\, dS + (p - p_{eq})\, dV,$$

which measures the degree of irreversibility for the former process.

3.6. Equilibrium Conditions and the Sense of Natural Processes

Let us consider again the fundamental Equation (20):

$$T\, dS \geqslant dU + p\, dV$$

and let us assume that we have a closed system in a certain state. We may imagine that this system undergoes an infinitesimal change, through an arbitrarily chosen process. We call this a *virtual* change, and the process that produces it may in principle be a natural, impossible or reversible one. To find out which is the case, we calculate the variations DS, DU and DV (where the differential symbol D rather than d indicates that the variation is a virtual and not a real one) and we try the previous equation with these values. If we obtain

$$T\, DS > DU + p\, DV$$

the process is a natural or spontaneous one. If an equality sign holds, the imagined process is a reversible one. If the inequality is of an opposite sense, the process is impossible.

It should be remarked that, although the process investigated may turn out to be of any kind (reversible, natural or impossible), the actual calculation of virtual variations of some of the state functions will require consideration of reversible paths.

We may now repeat this procedure with every process that we can imagine. If none of them are natural (if in every case we obtain signs $\leqslant$), the system is in thermodynamic equilibrium, as all virtual processes turn out to be either impossible or reversible. It may happen that for some of these processes signs $\leqslant$ are obtained, while for others we find $>$; the system is then in equilibrium with respect to the former, but not to the latter ones. This would be the case, for instance, of ice at a temperature above $0\,°C$ and of water below $0\,°C$; we may have, in both cases, thermal and mechanical equilibrium, but not chemical equilibrium, as the change of physical state will be a spontaneous process. The ice will actually be melting; the water may remain indefinitely liquid, but this is a *metastable* equilibrium (cf. Chapter 1, Section 4).

It will be convenient, in general, to apply one or another of Equations (26), imposing *restrictive conditions* to the virtual processes (e.g., that they occur at constant temperature and pressure). By so doing, we obtain particularly simple conditions (e.g., for constant T and p, we should have $DG \geqslant 0$ for every virtual process, if the system is in equilibrium). These are summarized below:

General equation:		$T dS \geqslant \delta Q$ or any of Equations (26)			
Restrictive conditions:	Adiabatic process $\delta Q = 0$	Isentropic process $dS = 0$		Isothermal process $dT = 0$	
		Isochoric $dV = 0$	Isobaric $dp = 0$	Isochoric $dV = 0$	Isobaric $dp = 0$
Resulting equation:	$dS \geqslant 0$	$dU \leqslant 0$ (from 20)	$dH \leqslant 0$ (from 21)	$dF \leqslant 0$ (from 24)	$dG \leqslant 0$ (from 25)
Equilibrium criterion:	$DS \leqslant 0$	$DU \geqslant 0$	$DH \geqslant 0$	$DF \geqslant 0$	$DG \geqslant 0$

Thus an important use of the thermodynamic potentials, whose differentials are expressed in terms of the differentials of T, p, V, is to obtain simple criteria for the natural sense of processes and of thermodynamic equilibrium, convenient for practical application.

The equations in the table tell us that, if a system is in equilibrium, the function S is a maximum with respect to any adiabatic change, and that the functions U, H, F, G are minima with respect to isentropic isochoric, isentropic isobaric, isothermal isochoric and isothermal isobaric changes, respectively. There also becomes apparent the convenience of associating U with S, V, etc.

A simple example will serve to illustrate these ideas. Let us consider a system consisting of water in the presence of water vapor at pressure p, and let the virtual process be the condensation of an infinitesimal number n of moles of vapor. It will be convenient to use the Gibbs function G. In order to calculate DG, we must perform the virtual change along a reversible path; this can be the isothermal compression of n moles of vapor from p to the equilibrium pressure p_s (saturated vapor pressure), and then condensation at constant T and p_s. In the first step we have:

$$D_1 G = n \int_p^{p_s} V \, dp = n \int_p^{p_s} R^* T \, d \ln p = n R^* T \ln \frac{p_s}{p}.$$

During the second step, $D_2 G = 0$, because both T and p_s remain constant. We then bring back the condensed n moles to the original pressure p. This contributes a term

$$D_3 G = n \int_p^{p_s} V_w \, dp \cong n V_w (p - p_s),$$

(where V_w is the molar volume of water) which has opposite sign to the previous term D_1G but is much smaller, because $V_w \ll V$ (molar volume of vapor); in fact it is quite negligible. Therefore*:

$$DG = D_1G + D_3G > 0 \quad \text{if } p < p_s$$

$$DG = D_1G + D_3G < 0 \quad \text{if } p > p_s.$$

In the first case, the process is impossible; in the second one, it is spontaneous. If we reverse the virtual process, the opposite will be true. We can conclude that only if $p_s = p$ (for which $DG = 0$ in both cases) can there be equilibrium. We can also conclude that no other equilibrium vapor pressure exists, because only for $p = p_s$ can DG vanish (reversible process).

3.7. Calculation of Entropy

In order to have an expression for dS with coefficients depending on directly measurable properties, so that we may integrate it to obtain values for finite differences ΔS, we start by expressing dS as a function of T and p as independent variables:

$$dS = \left(\frac{\partial S}{\partial T}\right)_p dT + \left(\frac{\partial S}{\partial p}\right)_T dp. \tag{34}$$

This may be compared with the joint expression for both principles that uses enthalpy (Equation (21)):

$$dS = \frac{1}{T} dH - \frac{V}{T} dp \tag{35}$$

and developing dH as a function of T and p:

$$dS = \frac{1}{T}\left(\frac{\partial H}{\partial T}\right)_p dT + \frac{1}{T}\left[\left(\frac{\partial H}{\partial p}\right)_T - V\right] dp. \tag{36}$$

Comparison of the coefficients of dT in Equations (34) and (36) shows (dT and dp being independent, we may compare for $dp = 0$):

$$\left(\frac{\partial S}{\partial T}\right)_p = \frac{1}{T}\left(\frac{\partial H}{\partial T}\right)_p = \frac{C_p}{T} \tag{37}$$

(cf. Chapter II, Section 4).

---

* This can be demonstrated simply in the critical case of p_s approximately equal to p, for which D_1G becomes, on expansion of the natural logarithm,

$$D_1G = nR^*T \ln(p_s/p) = nVp \ln\left(1 + \frac{p_s - p}{p}\right)$$

$$= nV\{(p_s - p) + \ldots\}.$$

Hence
$$DG = n(p_s - p)(V - V_w) + \ldots$$

The variation of S with p at constant T is given by one of Maxwell's relations (Equations (30)):

$$\left(\frac{\partial S}{\partial p}\right)_T = -\left(\frac{\partial V}{\partial T}\right)_p. \tag{38}$$

This derivative is directly measurable, and is usually expressed as the isobaric expansion coefficient $(1/V)(\partial V/\partial T)_p = \alpha_p$.

We have thus the expression

$$dS = \frac{C_p}{T} dT - \left(\frac{\partial V}{\partial T}\right)_p dp = C_p \, d\ln T - V\alpha_p \, dp \tag{39}$$

that can be applied conveniently to the computation of entropy changes. A similar expression can be derived as a function of T and V, by introducing U instead of H (therefore using Equation (20) rather than (21)) and following a similar derivation. This gives

$$dS = \frac{C_v}{T} dT + \left(\frac{\partial p}{\partial T}\right)_V dV = C_v \, d\ln T + p\alpha_v \, dV \tag{40}$$

where $\alpha_v = (1/p)(\partial p/\partial T)_v$ is the isochoric coefficient of pressure rise with temperature.

Equations (39) and (40) permit, by integration from experimental data, the direct calculation of the entropy changes associated with changes in the state of a homogeneous system of constant chemical composition, when no changes in the physical state (fusion, vaporization, sublimation) take place. When the latter changes occur the variation in the entropy is easily calculated; if the change of phase occurs reversibly, the pressure and the temperature are constant during the process (as will be discussed in the next chapter) and $\Delta H = L$. Therefore the entropy changes by

$$\Delta S = \frac{\Delta H}{T} = \frac{L}{T} \tag{41}$$

(for 1 mol). By using both (39), or (40), and (41) the entropy of a substance can be referred to any appropriate reference state.

3.8. Thermodynamic Equations of State. Calculation of Internal Energy and Enthalpy

By comparing the coefficients of dp in Equations (34) and (36), and introducing Equation (38), we see that

$$\left(\frac{\partial H}{\partial p}\right)_T = V - T\left(\frac{\partial V}{\partial T}\right)_p = V(1 - T\alpha_p). \tag{42}$$

By using U instead of H, and T, V instead of T, p, a similar derivation gives

$$\left(\frac{\partial U}{\partial V}\right)_T = T\left(\frac{\partial p}{\partial T}\right)_V - p = p(T\alpha_v - 1). \tag{43}$$

Equations (42) and (43) are called the *thermodynamic equations of state*. They solve a problem remaining from Chapter II, Section 5: that of calculating the variations in U and H for processes in which V and p vary. Thus by writing the total differentials

$$dU = \left(\frac{\partial U}{\partial T}\right)_V dT + \left(\frac{\partial U}{\partial V}\right)_T dV \tag{44}$$

$$dH = \left(\frac{\partial H}{\partial T}\right)_p dT + \left(\frac{\partial H}{\partial p}\right)_T dp \tag{45}$$

and introducing Chapter II, Equations (29) and (30) and Equations (42) and (43), we obtain

$$dU = C_v\, dT + p(T\alpha_v - 1)\, dV \tag{46}$$

$$dH = C_p\, dT + V(1 - T\alpha_p)\, dp \tag{47}$$

The integration of these formulas completely solves the problem of integrating internal energy and enthalpy for processes without changes of phase. Obviously, the last terms in both (46) and (47) vanish for ideal gases.

If changes of phase occur at constant pressure and temperature, the change in enthalpy, per mole, is given directly by

$$\Delta H = L \tag{48}$$

and that of internal energy, by

$$\Delta U = \Delta H - p\, \Delta V = L - p\, \Delta V \tag{49}$$

The last term in (49) will be negligible for fusion, because the molar volumes of condensed phases are small. For vaporization and sublimation, $\Delta V \cong V_v =$ molar volume of the vapor, and if the vapor is approximated by an ideal gas,

$$\Delta U = L - R^*T \tag{50}$$

By the use of formulas (46) to (50), the internal energy and the enthalpy of a substance can be referred to any appropriate reference state.

3.9. Thermodynamic Functions of Ideal Gases

Either by introducing the gas law into Equation (39), which is general for any system, or by starting from $dS = \delta Q/T$ and substituting for δQ the expressions of the first principle for ideal gases (Chapter II, Equations (39) and (40)), we derive for ideal gases

$$dS = C_p \, d\ln T - R^* \, d\ln p$$
$$= C_v \, d\ln T + R^* \, d\ln V \qquad (51)$$
$$= C_v \, d\ln p + C_p \, d\ln V$$

where we have used the relation $d\ln T = d\ln p + d\ln V$ (logarithmic differentiation of the gas law).

Integration gives:

$$S = C_p \ln T - R^* \ln p + b$$
$$= C_v \ln T + R^* \ln V + b' \qquad (52)$$
$$= C_v \ln p + C_p \ln V + b''.$$

And by using the gas law and the relation $C_p - C_v = R^*$ it is easily seen that $b' = b - R^* \ln R^*$ and $b'' = b - C_p \ln R^*$.

With the definition of potential temperature (Chapter II, Equation (55)), we have

$$d\ln\theta = d\ln T - \varkappa \, d\ln p \qquad (53)$$

and for the entropy

$$dS = C_p \, d\ln\theta. \qquad (54)$$

Taking C_p as constant, θ may be thus considered as an alternative variable for entropy in reversible processes.

Having now the expressions of U, H (Chapter II, Equations (37) and (38)) and S for ideal gases, those for F and G can immediately be written down. For instance, we have for G:

$$G = (C_p T - C_p T \ln T - bT) + R^* T \ln p + a \qquad (55)$$

as a function of T and p, where the terms dependent on T alone are between brackets and a is the additive constant for H.

3.10. Entropy of Mixing for Ideal Gases

Let us assume that we have a mixture of two ideal gases, with partial pressures p_1 and p_2, at temperature T and occupying a volume V. We want to know what difference of entropy exists between this state and that in which the two gases are separate, at their corresponding partial pressures. In order to calculate ΔS, we must link both states by a reversible process; this is done by performing the mixture with the ideal experiment illustrated in Figure III-2. Two cylinders of equal volume V contain initially the two gases; both cylinders have a semipermeable membrane at one end and can be inserted one into the other, as indicated in the figure (the wall's thickness is neglected). A (fixed to cylinder 2) is a semipermeable membrane only permeable to gas 1, and B (fixed to cylinder 1) is a semipermeable membrane only permeable to

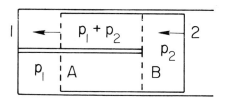

Fig. III-2. Isentropic mixing of ideal gases.

gas 2. As the gas that diffuses freely through each membrane does not exert any pressure on it, the device is in equilibrium and the mixture can be performed reversibly (using an infinitesimal excess of pressure on the external bases until both cylinders enclose the same volume) and isothermally. Neither gas performs any work, and their internal energies remain constant (as the temperature remains constant). Therefore, according to the first principle the process is also adiabatic and, being reversible, isentropic. We conclude that $\Delta S = 0$. We can say that the entropy of the mixture is equal to the sum of the *partial entropies*, defined as the entropies of the gases at their partial pressures, when pure and occupying the same volume at the same temperature. This conclusion can obviously be extended to any number of gases, and is known as Gibbs' theorem.

The process of mixing just described must not be confused with that of two gases at the same pressure mixing by diffusion into each other, with a final volume equal to the sum of the two partial volumes. This is an irreversible process, with a finite increase in total entropy.

3.11. Difference Between Heat Capacities at Constant Pressure and at Constant Volume

In Chapter II, Section 5, we found that $C_p - C_v = R^*$ for ideal gases. With the results of the previous sections, we are now in a position to derive the general expression of $C_p - C_v$ for any homogeneous system of constant composition.

The following well-known relation between partial derivatives can be readily obtained

$$\left(\frac{\partial U}{\partial T}\right)_p = \left(\frac{\partial U}{\partial V}\right)_T \left(\frac{\partial V}{\partial T}\right)_p + \left(\frac{\partial U}{\partial T}\right)_V. \tag{56}$$

From Chapter II, Equations (29) and (30) we see that the last term of Equation (56) is C_v, and that we can write for C_p, considering that $H = U + pV$,

$$C_p = \left(\frac{\partial H}{\partial T}\right)_p = \left(\frac{\partial U}{\partial T}\right)_p + p\left(\frac{\partial V}{\partial T}\right)_p. \tag{57}$$

Solving for $(\partial U/\partial T)_p$ in Equation (57), introducing this expression into Equation (56) and rearranging we obtain:

$$C_p - C_v = \left[p + \left(\frac{\partial U}{\partial V}\right)_T\right]\left(\frac{\partial V}{\partial T}\right)_p. \tag{58}$$

By a similar procedure, using H instead of U, we can also obtain the expression

$$C_p - C_v = \left[V - \left(\frac{\partial H}{\partial p}\right)_T\right]\left(\frac{\partial p}{\partial T}\right)_V. \tag{59}$$

Introducing now Equation (43) into (58), we obtain

$$C_p - C_v = T\left(\frac{\partial p}{\partial T}\right)_V\left(\frac{\partial V}{\partial T}\right)_p = TpV\alpha_v\alpha_p. \tag{60}$$

The coefficients α_v, α_p are small for condensed phases (solid or liquid), and therefore the difference $C_p - C_v$ is usually (but not always) small, and can be neglected in a first approximation.

We can express the calculated difference using the coefficient of compressibility $k = -(1/V)(\partial V/\partial p)_T$ instead of α_V. We apply the following relation between partial derivatives

$$\left(\frac{\partial V}{\partial p}\right)_T\left(\frac{\partial p}{\partial T}\right)_V\left(\frac{\partial T}{\partial V}\right)_p = \frac{(\partial V/\partial p)_T(\partial p/\partial T)_V}{(\partial V/\partial T)_p} = -1 \tag{61}$$

which gives

$$\frac{\alpha_p}{\alpha_v k} = p \tag{62}$$

so that Equation (60) can also be written

$$C_p - C_v = \frac{TV\alpha_p^2}{k}. \tag{63}$$

These formulas refer the difference $C_p - C_v$ to quantities that are directly measurable.

Application of Equation (63) to *water* gives

$$C_p - C_v = 0.13 \text{ cal mol}^{-1}\text{K}^{-1} \text{ at } 0\,°\text{C}$$

$$= 0 \quad \text{cal mol}^{-1}\text{K}^{-1} \text{ at } 4\,°\text{C}$$

(temperature of density maximum)

$$= 0.18 \text{ cal mol}^{-1}\text{K}^{-1} \text{ at } 25\,°\text{C}$$

to be compared with

$$C_p = 18.02 \text{ cal mol}^{-1}\text{K}^{-1}.$$

Therefore, in most problems the difference can be neglected with sufficient approximation. This is not always the case for liquids; for instance, for *ethyl ether*

$$C_p - C_v = 11 \text{ cal mol}^{-1}\text{K}^{-1}$$

to be compared with

$$C_p = 40 \text{ cal mol}^{-1}\text{K}^{-1}; \quad C_v = 29 \text{ cal mol}^{-1}\text{K}^{-1}.$$

We shall only be concerned with water and ice, for which the approximation $C_p \cong C_v$ is valid.

The specific heat capacities of water c_w and ice c_i are

$$c_w = 1.00 \text{ cal g}^{-1}\text{K}^{-1} = 4.18 \times 10^3 \text{ J kg}^{-1}\text{K}^{-1} \text{ at } 40\,°\text{C}$$
$$1.01 \text{ cal g}^{-1}\text{K}^{-1} = 4.22 \times 10^3 \text{ J kg}^{-1}\text{K}^{-1} \text{ at } 0\,°\text{C}$$
$$1.14 \text{ cal g}^{-1}\text{K}^{-1} = 4.77 \times 10^3 \text{ J kg}^{-1}\text{K}^{-1} \text{ at } -40\,°\text{C}$$
$$c_i = 0.504 \text{ cal g}^{-1}\text{K}^{-1} = 2.11 \times 10^3 \text{ J kg}^{-1}\text{K}^{-1} \text{ at } 0\,°\text{C}$$
$$0.43 \text{ cal g}^{-1}\text{K}^{-1} = 1.81 \times 10^3 \text{ J kg}^{-1}\text{K}^{-1} \text{ at } -40\,°\text{C}$$

PROBLEMS

1. One mole of ideal gas occupied 10 l at 300 K. It then expands into a vacuum until its volume is 20 l. Calculate the change in entropy and in the Gibbs function.
2. Calculate ΔS (where it is possible) and $\delta Q/T$ for the four processes mentioned in problem Chapter II, Section I, all resulting in the same isothermal compression of an ideal gas.
3. 0.01 kg of supercooled water at $-10\,°\text{C}$ is observed to freeze. The latent heat of fusion is lost to the surroundings and the ice regains the previous temperature. Calculate $\Delta U, \Delta H, \Delta S$ and ΔG. Neglect the volume variations and the effect of pressure changes in condensed phases and show that within this approximation, $l_s = l_v + l_f$ at the temperature stated. Under what conditions is this strictly true?
4. Assume that only work of compression can be done on a system. What criteria can be derived from the second principle to decide if the system is in equilibrium with respect to (a) adiabatic changes, (b) isothermal changes, (c) isothermal isobaric changes?
5. In problem 5 from Chapter II the effect of pressure change was neglected. Compute now what is the effect of increasing the pressure from $e_{w,0}$ to $e_{w,20}$ (saturation vapor pressures at 0 and 20 °C) on the internal energy and on the enthalpy of 1 g of water at 0 °C. Compare these values with those of problem 5 from Chapter II. The cubic expansion coefficient of water at constant pressure is $-6 \times 10^{-5} \text{ K}^{-1}$ at 0 °C.
6. Compute the variation in specific enthalpy and in specific entropy of liquid water at 0 °C
   (a) when the pressure is decreased from 1 atm to 0.5 atm.
   (b) when the temperature is decreased to $-10\,°\text{C}$ (the water remaining liquid).
   Coefficient of cubic expansion of water at 0 °C and constant pressure:
   $\alpha_p = -6 \times 10^{-5} \text{ K}^{-1}$.

7. How will the air pressure change, if the entropy decreases by 10 J kg^{-1} K^{-1} and the air temperature increases by 2%?
8. In what direction (increase or decrease) does the enthalpy of a system vary when undergoing an irreversible isentropic and isobaric process?
9. 1 kg of dry air at 300 K and 1000 mb is brought to 235 K and a pressure p.
   (a) Knowing that the specific entropy increased by 100 J kg^{-1} K^{-1}, derive the value of p.
   (b) What is the final potential temperature Θ_2?
10. Supercooled water droplets in a cloud freeze. After a while, they have recovered the initial temperature. The pressure did not change. Is the difference of free enthalpy (Gibbs' free energy) between the initial and final states zero, positive or negative? Explain.
11. 1 g of water at 0 °C is cooled to $-10\,°C$, then freezes (the latent heat of freezing being rapidly removed by ventilation during the process) and the ice is finally brought back to 0 °C. Is the total entropy change equal to (i) $-l_{f,-10}/263.15$? (ii) to $(c_i - c_w)\ln\dfrac{273.15}{263.15} - l_{f,-10}/263.15$? (iii) to $-l_{f,0}/273.15$? (c_i, c_w = specific heat capacities of ice and water, per gram; $l_{f,-10}$, $l_{f,0}$ = latent heat of freezing at $-10\,°C$ and at 0 °C, per gram.) Explain.

CHAPTER IV

WATER-AIR SYSTEMS

4.1. Heterogeneous Systems

The fundamental equations seen in the previous chapter (Chapter III, Equation (26)) are valid for closed systems; no assumption regarding the internal structure of the system is implicit in them. Any of the four equivalent equations is a joint expression of the first and second principles, and therefore contains all that thermodynamics can say about closed systems, except as regards the third principle. Furthermore, while we deal with systems that, besides being closed, are homogeneous and of constant chemical composition*, it is not necessary to specify how the thermodynamic functions depend on the composition of the system, and in order to determine its state only two independent variables must be known (e.g., T and p); we do not need to take into account as another independent variable the total mass or the total number of moles, because this is assumed to be constant and, if we know the values of the extensive functions for the unit mass or the mole, generalization to any mass is done by simply multiplying by m or n (n: number of moles) (cf. Chapter I, Section 3).

However, in the study of heterogeneous systems, we are concerned with the conditions of internal equilibrium between the phases. And even if we consider a closed heterogeneous system, each one of its phases constitutes in its turn a homogeneous system (a 'subsystem'), which will be open, as we shall admit the possibility of exchange of components between phases. We must find therefore the form of the characteristic functions and of the fundamental equations, first for open homogeneous systems, and then for the total heterogeneous system.

The composition of an open homogeneous system is not constant in general; therefore, besides the two independent variables we used for closed systems (e.g., T and p) we have to consider the masses or concentrations of all the components, rather than a single mass m. In order to do so, we must specify precisely the notion of *number of components*; this is defined as the minimum number of chemical substances with which the composition of all phases can be expressed, separately for each one of them. In other words, the masses of the components can be considered as mathematical variables with which the composition of each phase can be expressed; the total number of such variables (chosen in such a way as to be a minimum) is defined as the number of components.

---

* When we say 'chemical composition', we refer not only to the chemical species involved but also to their physical states.

We shall restrict ourselves to the specific case in which we are interested while introducing the new concepts and formulas. To develop the expressions for the most general case would require writing somewhat more complicated formulas, and this generality is unnecessary for the relatively simple systems that we have to study. In a later section (Section 4) we shall summarize the basic equations and the formulas will be written in their general way. Consideration of chemical reactions will of course be excluded, as unnecessary for our applications.

We are concerned with a system composed of moist air and water in either one of its two condensed states: liquid water or ice. The system composed of humid air, liquid water *and* ice is of considerable interest in cloud physics, but in that case it is not in general a system in equilibrium and it would therefore be a subject for the thermodynamics of irreversible processes; from the point of view of equilibrium conditions, its study would be restricted to that of the triple point of water in the presence of air.

Our system has two phases: one condensed and one gaseous phase. Obviously, water substance may be chosen as one of the components. If we should consider that the different gases do not dissolve in water in the same proportion as they are in air, each gas would be a different component; but the solubility of the air gases in water or ice is of no significance to our purposes, and we shall neglect it from the beginning. We may then count the constant mixture which we have called "dry air" as the second and last component of the system, and we shall consider that this component is restricted to the gas phase. On the other hand, the two phases can exchange water component: they constitute thus open systems.

The amounts of components will be expressed by their number of moles, n_d, n_v, n_c, n_g, n_w, n_i, where the subscripts d, v, c, g, w, i stand for dry air, water vapor, water in condensed phase, gaseous phase, liquid water and ice (n_c may be either n_w or n_i), respectively. Alternatively, they may be expressed by their masses m, with similar subindices.

Let Z be any extensive property (such as U, S, G, V, etc.). We shall call $Z_{g,tot}$, $Z_{c,tot}$ and Z_{tot} the total values of Z for the gas phase, the condensed phase and the total system, respectively, so that

$$Z_{tot} = Z_{g,tot} + Z_{c,tot} \tag{1}$$

Z_d, Z_v and Z_c, on the other hand, will be used as symbols of the molar values.

For each phase, the total value of Z is no longer a function of only two variables (e.g., p and T), because the mass of its components may also vary. The total differentials of Z for these two open systems can be expressed:

$$dZ_{g,tot} = \left(\frac{\partial Z_{g,tot}}{\partial T}\right)_{p,n} dT + \left(\frac{\partial Z_{g,tot}}{\partial p}\right)_{T,n} dp + \left(\frac{\partial Z_{g,tot}}{\partial n_d}\right)_{T,p,n} dn_d +$$

$$+ \left(\frac{\partial Z_{g,tot}}{\partial n_v}\right)_{T,p,n} dn_v \tag{2}$$

$$dZ_{c,tot} = \left(\frac{\partial Z_{c,tot}}{\partial T}\right)_{p,n} dT + \left(\frac{\partial Z_{c,tot}}{\partial p}\right)_{T,n} dp + \left(\frac{\partial Z_{c,tot}}{\partial n_c}\right)_{T,p} dn_c \qquad (3)$$

where n as a subscript means 'at constant composition' (that is, keeping constant all values of n_k in the system, except the one that is being varied, in the last derivatives).

By a general definition,

$$\bar{Z}_k = \left(\frac{\partial Z_{v,tot}}{\partial n_k}\right)_{T,p,n} \qquad (4)$$

is called the *partial molar property* Z of the component k in the phase considered (k stands for any component and v for any phase).

Similarly,

$$\bar{z}_k = \left(\frac{\partial Z_{v,tot}}{\partial m_k}\right)_{T,p,m} \qquad (5)$$

is called the *partial specific property* Z, the mass subscript m having a similar meaning as n in the previous definition.

If in particular $Z = G$, the partial molar Gibbs function is written

$$\bar{G}_k = \mu_k \qquad (6)$$

and is called the *chemical potential* of component k. It may be worth noting that partial molar properties are always referred to the independent variables T and p, as defined above, whatever the meaning of Z; G is in particular the function that is usually associated with T and p (cf. Chapter III, Section 5).

The interpretation of these quantities becomes clearer by considering for one phase a process at constant T and p; taking the gas phase, for example, we have

$$dZ_{g,tot} = \bar{Z}_d \, dn_d + \bar{Z}_v \, dn_v. \qquad (7)$$

The partial molar properties are in general dependent on the composition of the phase. If we integrate this expression at constant composition, we obtain ($\bar{Z}_d$ and $\bar{Z}_v$ being then constants)

$$Z_{g,tot} = n_d \bar{Z}_d + n_v \bar{Z}_v. \qquad (8)$$

Physically, we may imagine this integration as the process of adding simultaneously both components in a constant proportion until the total mass of gaseous phase is obtained. The partial molar property $\bar{Z}_k$ is therefore the amount contributed by one mole of component k when added while maintaining a constant proportion of all components. Equation (8) is of course of general validity, independent of the fact that it was derived from an argument based on a process at constant T, p and composition; it could not be otherwise, as $Z_{g,tot}$ and the $\bar{Z}_i$ are all state functions.

By differentiating Equation (8) and comparing with Equation (7), we find that

$$n_d \, d\bar{Z}_d + n_v \, d\bar{Z}_v = 0. \qquad (9)$$

This is called the Gibbs-Duhem equation, valid for a system at constant T and p, as applied to our particular example.

The partial molar property $\bar{Z}_k$ will, in general, differ from the molar value Z_k for the pure component, but it can be taken as equal with very good approximation in our system. Therefore, we write

$$\bar{Z}_d \cong Z_d \quad \text{or} \quad \bar{z}_d \cong z_d \tag{10}$$

and

$$\bar{Z}_v \cong Z_v \quad \text{or} \quad \bar{z}_v \cong z_v.$$

For the condensed phase, as we have only one component, obviously $\bar{Z}_c \equiv Z_c$ (or $\bar{z}_c \equiv z_c$), and $Z_{c,\,tot} = n_c Z_c = m_c z_c$. In particular, we can write for the Gibbs function:

$$\mu_d \cong G_d, \quad \mu_v \cong G_v, \quad \mu_c \equiv G_c.$$

Introducing Equations (4), (8) and (10) into Equations (2) and (3), we may write

$$dZ_{g,\,tot} = \left(\frac{\partial Z_{g,\,tot}}{\partial T}\right)_{p,n} dT + \left(\frac{\partial Z_{g,\,tot}}{\partial p}\right)_{T,n} dp + Z_d \, dn_d + Z_v \, dn_v$$

$$= n_d \left(\frac{\partial Z_d}{\partial T}\right)_{p,n} dT + n_v \left(\frac{\partial Z_v}{\partial T}\right)_{p,n} dT + n_d \left(\frac{\partial Z_d}{\partial p}\right)_{T,n} dp +$$

$$+ n_v \left(\frac{\partial Z_v}{\partial p}\right)_{T,n} dp + Z_d \, dn_d + Z_v \, dn_v \tag{11}$$

$$dZ_{c,\,tot} = \left(\frac{\partial Z_{c,\,tot}}{\partial T}\right)_{p,n} dT + \left(\frac{\partial Z_{c,\,tot}}{\partial p}\right)_{T,n} dp + Z_c \, dn_c$$

$$= n_c \left(\frac{\partial Z_c}{\partial T}\right)_{p,n} dT + n_c \left(\frac{\partial Z_c}{\partial p}\right)_{T,n} dp + Z_c \, dn_c. \tag{12}$$

At this stage we may introduce the condition, which will always be assumed, that the total system is a closed one. This is expressed in our case by

$$n_d = \text{const.} \tag{13}$$

$$n_v + n_c = n_t = \text{const.}$$

or by

$$dn_d = 0$$

$$dn_v = -dn_c. \tag{14}$$

If we introduce these conditions in the previous expressions, and add them to obtain dZ_{tot} for the whole system, we have:

$$dZ_{tot} = \left(\frac{\partial Z_{tot}}{\partial T}\right)_{p,n} dT + \left(\frac{\partial Z_{tot}}{\partial p}\right)_{T,n} dp + (Z_v - Z_c) \, dn_v \tag{15}$$

where
$$Z_{\text{tot}} = n_d Z_d + n_v Z_v + n_c Z_c$$

$$\left(\frac{\partial Z_{\text{tot}}}{\partial T}\right)_{p,n} = n_d \left(\frac{\partial Z_d}{\partial T}\right)_{p,n} + n_v \left(\frac{\partial Z_v}{\partial T}\right)_{p,n} + n_c \left(\frac{\partial Z_c}{\partial T}\right)_{p,n}$$

$$\left(\frac{\partial Z_{\text{tot}}}{\partial p}\right)_{T,n} = n_d \left(\frac{\partial Z_d}{\partial p}\right)_{T,n} + n_v \left(\frac{\partial Z_v}{\partial p}\right)_{T,n} + n_c \left(\frac{\partial Z_c}{\partial p}\right)_{T,n}.$$

If masses are used instead of numbers of moles, the expression will be written

$$dZ_{\text{tot}} = \left(\frac{\partial Z_{\text{tot}}}{\partial T}\right)_{p,m} dT + \left(\frac{\partial Z_{\text{tot}}}{\partial p}\right)_{T,m} dp + (z_v - z_c) dm_v \tag{16}$$

with
$$Z_{\text{tot}} = m_d z_d + m_v z_v + m_c z_c$$

$$\left(\frac{\partial Z_{\text{tot}}}{\partial T}\right)_{p,m} = m_d \left(\frac{\partial z_d}{\partial T}\right)_{p,m} + m_v \left(\frac{\partial z_v}{\partial T}\right)_{p,m} + m_c \left(\frac{\partial z_c}{\partial T}\right)_{p,m}$$

$$\left(\frac{\partial Z_{\text{tot}}}{\partial p}\right)_{T,m} = m_d \left(\frac{\partial z_d}{\partial p}\right)_{T,m} + m_v \left(\frac{\partial z_v}{\partial p}\right)_{T,m} + m_c \left(\frac{\partial z_c}{\partial p}\right)_{T,m}.$$

Z_{tot} can be written as mz, m being the total mass and z the average specific value of Z.

4.2. Fundamental Equations for Open Systems

Let us consider now the Gibbs function for the gas phase. Equation (2), with the definitions (4) and (6) and $Z = G$, gives

$$dG_{g,\text{tot}} = \left(\frac{\partial G_{g,\text{tot}}}{\partial T}\right)_{p,n} dT + \left(\frac{\partial G_{g,\text{tot}}}{\partial p}\right)_{T,n} dp + \mu_d \, dn_d + \mu_v \, dn_v \tag{17}$$

where
$$G_{g,\text{tot}} = n_d \bar{G}_d + n_v \bar{G}_v = n_d \mu_d + n_v \mu_v. \tag{18}$$

We can consider a reversible process without exchange of mass (system acting as closed: $dn_d = dn_v = 0$) and compare Equation (17) with Chapter III, Equation (25):

$$dG_{g,\text{tot}} = -S_{g,\text{tot}} \, dT + V_{g,\text{tot}} \, dp. \tag{19}$$

As T and p are independent variables, the coefficients of dT and dp must be identical in both expressions; i.e.:

$$\left(\frac{\partial G_{g,\text{tot}}}{\partial T}\right)_{p,n} = -S_{g,\text{tot}} \tag{20}$$

$$\left(\frac{\partial G_{g,\text{tot}}}{\partial p}\right)_{T,n} = V_{g,\text{tot}} \tag{21}$$

where

$$S_{g,\text{tot}} = n_d \bar{S}_d + n_v \bar{S}_v$$

and $V_{g,\text{tot}}$ is given by a similar expression.

These equalities will still be valid for processes with mass exchange (the two partial derivatives being state functions themselves, therefore independent of the process we choose to consider); therefore we introduce them into Equation (17) and obtain, for an open phase:

$$dG_{g,\text{tot}} = -S_{g,\text{tot}}\, dT + V_{g,\text{tot}}\, dp + \mu_d\, dn_d + \mu_v\, dn_v. \tag{22}$$

This is the generalization of the fundamental equation (Chapter III, Equation (25)) for our particular open system and for processes occurring reversibly with respect to mechanical and thermal equilibrium. This last condition is implicit in the equality sign; otherwise the Equation (25) of Chapter III should have been taken with the inequality sign.

Introducing the equalities $G = U + pV - TS = H - TS = F + pV$ (cf. Chapter III, Section 5), three other equations could be obtained, similar to Chapter III, Equation (25) except for the additive terms $\mu_d\, dn_d + \mu_v\, dn_v$ that will appear in all of them. Similar expressions could be written for the condensed phase; in particular, for the Gibbs function we have

$$\begin{aligned} dG_{c,\text{tot}} &= \left(\frac{\partial G_{c,\text{tot}}}{\partial T}\right)_{p,n} dT + \left(\frac{\partial G_{c,\text{tot}}}{\partial p}\right)_{T,n} dp + \mu_c\, dn_c \\ &= -S_{c,\text{tot}}\, dT + V_{c,\text{tot}}\, dp + \mu_c\, dn_c \end{aligned} \tag{23}$$

where

$$S_{c,\text{tot}} = n_c S_c; \quad V_{c,\text{tot}} = n_c V_c; \quad G_{c,\text{tot}} = n_c G_c \equiv n_c \mu_c$$

4.3. Equations for the Heterogeneous System. Internal Equilibrium

We may assume now that we have both phases isolated and in equilibrium, both being at the same temperature and pressure. Let us bring them together. They shall continue to be in thermal and in mechanical equilibrium, because T and p are the same for both phases. But we do not know if they shall be in chemical equilibrium; for our particular system, this means that we do not know whether water (or ice, as the case may be) and vapor will remain in equilibrium, or whether condensation of vapor or evaporation of water (or sublimation of ice) will take place as a spontaneous process. The total value for dG is obtained by adding Equations (22) and (23):

$$dG_{\text{tot}} = -S_{\text{tot}}\, dT + V_{\text{tot}}\, dp + \mu_d\, dn_d + \mu_v\, dn_v + \mu_c\, dn_c. \tag{24}$$

If we now introduce the condition that the total heterogeneous system is closed, i.e., condition (14), we obtain

$$dG_{\text{tot}} = -S_{\text{tot}}\, dT + V_{\text{tot}}\, dp + (\mu_v - \mu_c)\, dn_v. \tag{25}$$

$G_{\text{tot}}, S_{\text{tot}}, V_{\text{tot}}$ are the total values for the system; e.g.:

$$G_{\text{tot}} = n_d \mu_d + n_v \mu_v + n_c \mu_c.$$

We consider now virtual displacements at constant T and p. The condition of equilibrium is (Chapter III, Section 6) $DG \geqslant 0$ for any arbitrary Dn_v. As Dn_v can be positive or negative, this condition requires that its coefficient vanish. Therefore, the condition of internal chemical equilibrium between the two phases is

$$\mu_v = \mu_c. \tag{26}$$

Again, introducing the relations between G and the other three characteristic functions U, H, F, a set of four equations of the type of Equation (24) (including (24)) could be obtained for the open heterogeneous system considered here, and another set of four equations like Equation (25) (including (25)) for the closed system.

4.4. Summary of Basic Formulas for Heterogeneous Systems *

The formulas considered so far have been developed for our particular system (water or ice + moist air). Their generalization to any heterogeneous system with c components and φ phases could be easily done without the need of any new concepts. The general expressions will now be written without derivation; the previous formulas will be easily recognized as particular cases.

The expression of the total differential of any extensive property Z is

$$dZ = \left(\frac{\partial Z}{\partial T}\right)_{p,n} dT + \left(\frac{\partial Z}{\partial p}\right)_{T,n} dp + \sum_{v=1}^{\varphi} \sum_{i=1}^{c} \bar{Z}_{iv} \, dn_{iv}. \tag{27}$$

We have now dropped the subscript 'tot'; Z refers to the total system. Subscripts i and v refer to the component and the phase, respectively; thus $\bar{Z}_{iv}$ means the partial molar Z property of component i in phase v.

For the particular case $Z = G$, we have

$$dG = \left(\frac{\partial G}{\partial T}\right)_{p,n} dT + \left(\frac{\partial G}{\partial p}\right)_{T,n} dp + \sum_{v=1}^{\varphi} \sum_{i=1}^{c} \mu_{iv} \, dn_{iv}. \tag{28}$$

If the system is closed, the conditions

$$\sum_{v=1}^{\varphi} dn_{iv} = 0 \quad (i = 1, 2, \ldots, c) \tag{29}$$

of conservation of components must hold, and Equation (28) becomes

$$dG = \left(\frac{\partial G}{\partial T}\right)_{p,n} dT + \left(\frac{\partial G}{\partial p}\right)_{T,n} dp + \sum_{v=2}^{\varphi} \sum_{i=1}^{c} (\mu_{iv} - \mu_{i1}) \, dn_{iv}. \tag{30}$$

The equivalent of Equation (30) for $Z = H$ would be:

$$dH = \left(\frac{\partial H}{\partial T}\right)_{p,n} dT + \left(\frac{\partial H}{\partial p}\right)_{T,n} dp + \sum_{v=2}^{\varphi} \sum_{i=1}^{c} (\bar{H}_{iv} - \bar{H}_{i1}) \, dn_{iv}. \tag{31}$$

---
* This section is intended both as a summary and a presentation of the formulas in their general form. The reader may omit it without loss of continuity.

Here the terms $(\bar{H}_{iv} - \bar{H}_{i1})$ have the meaning of molar latent heats for the passage of component i from phase 1 (arbitrarily chosen) to phase v.

The sum of Equations (11) and (12) would be a particular case of Equation (27). Equation (15) with $Z = G$ corresponds to Equation (30); Equation (17) corresponds to Equation (28).

Comparison with the fundamental equations (Chapter III, Equation (26)) and introduction of the other characteristic equations gives the following set of fundamental equations for processes in open heterogeneous systems occurring in conditions of thermal and mechanical reversibility:

$$dU = T\,dS - p\,dV + \sum_{v=1}^{\varphi} \sum_{i=1}^{c} \mu_{iv}\,dn_{iv}$$

$$dH = T\,dS + V\,dp + \sum_{v=1}^{\varphi} \sum_{i=1}^{c} \mu_{iv}\,dn_{iv}$$

$$dF = -S\,dT - p\,dV + \sum_{v=1}^{\varphi} \sum_{i=1}^{c} \mu_{iv}\,dn_{iv}$$

$$dG = -S\,dT + V\,dp + \sum_{v=1}^{\varphi} \sum_{i=1}^{c} \mu_{iv}\,dn_{iv}.$$

(32)

If the system is closed, the same formulas will hold with the substitution of

$$\sum_{v=2}^{\varphi} \sum_{i=1}^{c} (\mu_{iv} - \mu_{i1})\,dn_{iv} \tag{33}$$

for the double sum in Equation (32); here phase 1 is any arbitrarily chosen phase.

Equation (24) will be recognized as a particular case of the fourth (32). Equation (25) is the same with the substitution of the expression (33).

It may be shown from formulas (32) with (33) that the conditions of internal chemical equilibrium are

$$\mu_{i1} = \mu_{i2} = \ldots = \mu_{i\varphi} \quad (i = 1, 2, \ldots, c) \tag{34}$$

i.e., that the chemical potential of each component, be the same for all phases. Equation (26) was the particular case of Equations (34) for our system. The magnitude of the double sum gives a measure of the deviation from chemical equilibrium – therefore, of the irreversibility of the process.

The set of Equation (32) shows that four alternative definitions can be given of the chemical potential:

$$\mu_{iv} = \left(\frac{\partial U}{\partial n_{iv}}\right)_{S,V,n} = \left(\frac{\partial H}{\partial n_{iv}}\right)_{S,p,n} = \left(\frac{\partial F}{\partial n_{iv}}\right)_{T,V,n} = \left(\frac{\partial G}{\partial n_{iv}}\right)_{T,p,n}. \tag{35}$$

The last one was the definition used to introduce it (cf. Equation (6)). It should be noticed that the other three derivatives are *not* the partial molar properties (which are all defined at constant T and p).

4.5. Number of Independent Variables

We have seen that for closed systems of constant composition the number of independent variables whose values have to be known, in order to specify the state of the system, is two; for instance, T and p. If we now have a heterogeneous system of one component and two phases (e.g., water and water vapor), we have four variables corresponding to the two phases: p, T, p', T'. But if we impose the condition of equilibrium between the two phases, the four variables must obey three conditions:

$$p = p'$$
$$T = T' \tag{36}$$
$$G = G'$$

which reduces to one the number of independent variables. $G = \mu$ is here the molar Gibbs function. $G = G(T, p)$ and $G' = G(T', p')$, so that the third equation implies a relation between T, p and T', p'. Therefore, if we fix the temperature at which both phases are in equilibrium, the value of the pressure also becomes fixed, and vice versa. This defines curves $p = f(T)$ along which an equilibrium can exist – the equilibrium curves for changes of state, which we shall discuss for water substance.

If the three phases (solid, liquid, gas) are present simultaneously, we have three pairs of variables (pressure and temperature), but also three pairs of conditions for equilibrium:

$$p = p' = p''$$
$$T = T' = T'' \tag{37}$$
$$G = G' = G''.$$

Therefore, there is no independent variable; all values are fixed, and define what is called the *triple point*.

In our system of moist air plus one condensed phase of water, we may consider the same variables as for two phases of pure water (where the pressure of the gas phase would now become the partial pressure of water vapor), plus the partial pressure of dry air. As the number of conditions remain the same as before, we have now two independent variables, for instance T and p_d.

A general rule may be derived, which includes the previous systems as particular cases: the number of independent variables v that must be fixed in order to determine completely the equilibrium state of a heterogeneous system (its variance) is equal to the number of components c minus the number of phases φ plus two:

$$v = c - \varphi + 2. \tag{38}$$

This is the *phase rule*, derived by J. W. Gibbs. It is obtained immediately by a generalization of the previous argument (see Problem 12).

4.6. Phase-Transition Equilibria for Water

As we have seen above, phase-transition equilibria correspond to only one independent variable ($v=1$) and may be represented by curves $p=f(T)$, as shown schematically in Figure IV-1. They define in the p, T plane the regions where the water is in solid, liquid or gaseous state.

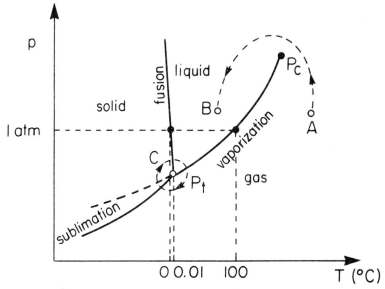

Fig. IV-1. Phase-transition equilibria for water.

The pressure of the vapor, when it is in equilibrium with the condensed phase at a certain temperature, is called the *vapor pressure* of the condensed phase at that temperature. The curves for sublimation and vaporization equilibria are thus the vapor pressure curves of ice and water. The curve for fusion corresponds to the equilibrium ice-water; its steep slope is negative, unlike what happens with similar curves for most other substances. The three curves meet at the triple point P_t, where the three phases coexist in equilibrium.

The extension of the vapor pressure curve of water at temperatures lower than the triple point, represented as a dashed curve, corresponds to supercooled water (metastable equilibrium).

The vapor pressure curve for water ends, for high temperatures, at the critical point P_c, beyond which there is no discontinuity between the liquid and gaseous phases. Thus, vapor represented by the point A could be transformed to water B without undergoing a two-phase condensation, if a path such as the one shown in Figure IV-1 is followed.

The pressure, temperature, and specific volumes for the three phases at the triple point are:

$$p_t = 610.7 \text{ Pa} = 6.107 \text{ mb}$$

$T_t = 273.16$ K

$v_{i,t} = 1.091 \times 10^{-3}$ m^3 kg^{-1}; ice density: $\varrho_{i,t} = 917$ kg m^{-3}

$v_{w,t} = 1.000 \times 10^{-3}$ m^3 kg^{-1}

$v_{v,t} = 206$ m^3 kg^{-1}.

The latent heat for changes of state, at 0 °C, are:

$l_v = 597.3$ cal g^{-1} = 2.5008×10^6 J kg^{-1}

$l_s = 677.0$ cal g^{-1} = 2.8345×10^6 J kg^{-1}

$l_f = 79.7$ cal g^{-1} = 0.3337×10^6 J kg^{-1}.

Within this approximation the same values hold for the triple point. It may be remarked that $l_s = l_f + l_v$. That this should be so is easily seen by considering a cycle around the triple point (C in Figure IV-1) performed by unit mass, and assuming that it tends to the point by becoming increasingly smaller. In the limit, the three transitions of state occur at constant temperature and pressure, so that for the cycle:

$$\Delta H = \Delta H_f + \Delta H_v - \Delta H_s = 0 \quad \text{at} \quad T_t, p_t,$$

or

$$l_f + l_v - l_s = 0.$$

The values for the critical point P_c (critical constants) are:

$T_c = 647$ K

$p_c = 218.8$ atm $= 2.22 \times 10^7$ Pa

$v_c = 3.07 \times 10^{-3}$ m^3 kg^{-1}.

The changes in pressure and specific volume along isotherms are summarized schematically in the diagram of Amagat-Andrews, shown in Figure IV-2. The curves are isotherms. At high temperatures, they tend to become equilateral hyperbolae, corresponding to ideal gas behavior in the water vapor. At lower temperatures, they become first deformed, until reaching a point of zero slope: the critical point P_c. Below this temperature, the vapor and liquid regions are separated by a zone of discontinuity, where liquid water and vapor coexist. Thus, if vapor represented by the point A is isothermically compressed, if follows the isotherm until reaching B. At that point, condensation starts, giving liquid corresponding to point C. As condensation proceeds, the mean specific volume becomes smaller, while pressure, as well as temperature, remains constant; the representative point slides along the horizontal line, from B to C. It reaches C when all vapor has condensed into liquid, and from there on it follows the compression curve of the liquid, which shows a much larger slope (smaller compressibility) than for the vapor.

The same type of process could be described for the sublimation region, below the temperature T_t. The straight line (isotherm) TT corresponds to the triple point.

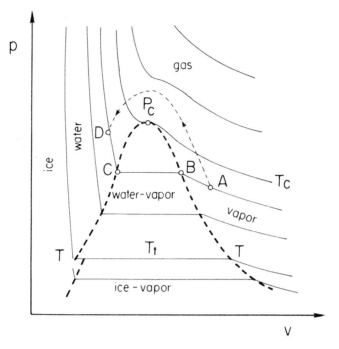

Fig. IV-2. Diagram of Amagat-Andrews.

Thus, several regions, corresponding to each of the three states, and to the changes of state, are defined in the diagram, as labelled in Figure IV-2. The region at temperatures higher than T_c is sometimes called the gas region as distinct from the vapor region (at the right, below T_c), to indicate that no two-phase discontinuity for condensation can occur in it.

As in Figure IV-1, a liquid state D can be reached from a vapor state A without discontinuity by a path such as shown by the dotted curve passing above P_c.

4.7. Thermodynamic Surface for Water Substance

Both diagrams p, T and p, v can be assembled into one tridimensional representation of the surface $f(p, v, T) = 0$, the equation of state for water substance in its three states. This is called the *thermodynamic surface* for water substance, and can be seen (schematically only) in Figure IV-3. The different regions are labelled in it, and the isotherms are drawn as full lines. A projection of the figure on the p, T plane would reproduce Figure IV-1. This means that the two-phase surfaces are all perpendicular to the p, T plane, as they can be conceived as made out of straight lines representing changes of state at constant p and T; the projection of these curves thus determines curves in the p, T plane, and the projection of the T_t isotherm is the triple point P_t. Similarly, the projection of Figure IV-3 on the p, v plane would reproduce Figure IV-2.

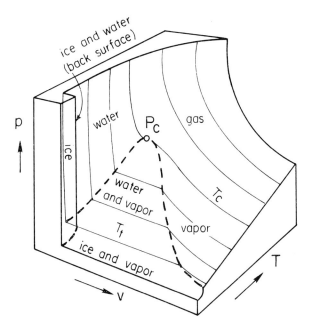

Fig. IV-3. Thermodynamic surface for water.

4.8. Clausius-Clapeyron Equation

We have seen that if we fix one of the two usual independent variables, let us say the pressure, and we heat a condensed phase, the temperature will increase until the equilibrium value is reached where two phases may coexist. The temperature then remains constant, as well as the pressure, until one of the phases has disappeared. For these changes we shall have

$$\Delta H = L \quad \text{or} \quad \Delta h = l \tag{39}$$

$$\Delta S = \frac{L}{T} \quad \text{or} \quad \Delta s = \frac{l}{T} \tag{40}$$

$$\Delta G = 0 \quad \text{or} \quad \Delta g = 0 \tag{41}$$

where L may be any of the three molar heats, and T the corresponding temperature of the change of phase.

We are now interested in calculating the relation between the changes in pressure and temperature when we change the conditions while preserving the equilibrium between the phases. At temperature T and the corresponding pressure p we have

$$G_a = G_b \tag{42}$$

a and b being the two phases. If we produce an infinitesimal change in the conditions, while preserving the equilibrium, we shall have, at $T+dT$ and $p+dp$,

$$G_a + dG_a = G_b + dG_b. \tag{43}$$

Therefore

$$dG_a = dG_b \tag{44}$$

or

$$-S_a\,dT + V_a\,dp = -S_b\,dT + V_b\,dp \tag{45}$$

and

$$\frac{dp}{dT} = \frac{\Delta S}{\Delta V} = \frac{\Delta H}{T\Delta V} = \frac{L}{T\Delta V}$$

or

$$\frac{dp}{dT} = \frac{l}{T\Delta v}. \tag{46}$$

Equation (46) gives us the ratio between the changes in the pressure and in the temperature along the equilibrium curves.

The physical meaning of Equation (46) may perhaps become clearer by considering a cycle for the particular case of vaporization. We shall have, in the p, v diagram, the cycle shown in Figure IV-4. We may go from water at T, p to vapor at $T+dT$, $p+dp$ by the two paths indicated in the diagram by the arrows. The changes in g in each step are also written in the diagram. Equating the variations in the specific Gibbs function for both equivalent processes (indicated in the figure for each of the steps), we obtain again Equation (46).

If we now apply Equation (46) to the melting of water, we obtain

$$\frac{dp}{dT} = -1.344 \times 10^5 \text{ mb K}^{-1}. \tag{47}$$

This shows that great increases in pressure correspond to small decreases in melting

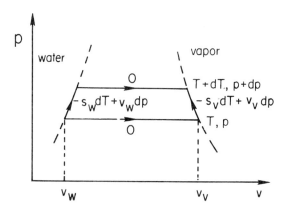

Fig. IV-4. Cycle related to the Clausius-Clapeyron equation.

temperature. The sign of the variation, peculiar to water and few other substances, corresponds to the fact that the solid contracts, instead of expanding, as it melts.

In the case of sublimation and vaporization, Equation (46) can be transformed and simplified by neglecting in Δv (or ΔV) the specific (or molar) volume of the condensed phase against that of the vapor ($\Delta v \cong v_v$), and by introducing the gas law to eliminate v_v:

$$\frac{dp}{dT} = \frac{lp}{R_v T^2} \tag{47}$$

or

$$\frac{d \ln p}{dT} = \frac{l}{R_v T^2} = \frac{L}{R^* T^2}. \tag{48}$$

This is the equation of Clausius-Clapeyron, where T is the sublimation or the boiling temperature, as the case may be. Its integration gives

$$\ln p = \frac{1}{R_v} \int l \frac{dT}{T^2} + \text{const.} = \frac{1}{R^*} \int L \frac{dT}{T^2} + \text{const.}$$

As a first approximation (particularly for small variations of T) l may be considered as constant, and

$$\ln p = -\frac{l}{R_v T} + \text{const.} = -\frac{L}{R^* T} + \text{const.} \tag{49}$$

or

$$\ln \frac{p_2}{p_1} = \frac{l}{R_v} \frac{\Delta T}{T_1 T_2} = \frac{L}{R^*} \frac{\Delta T}{T_1 T_2}. \tag{50}$$

l/R_v is thus given by the slope of the straight line $\ln p = f(1/T)$.

If a better approximation is desired, the expression in Chapter II, Equation (47) can be used for L. The integration gives then:

$$\ln p = \frac{1}{R^*} \left[-\frac{L_0}{T} + \Delta\alpha \ln T + \frac{\Delta\beta}{2} T + \frac{\Delta\gamma}{6} T^2 + \ldots \right] + \text{const.} \tag{51}$$

where the integration constant may be determined by an experimental pair of values T, p.

Let us consider vaporization. The first approximation (49), with the value of l_v above mentioned (Section 6), gives

$$\log p = 9.4041 - \frac{2354}{T} \tag{52}$$

where *log* is decimal logarithm, and p is given in mb.

The second approximation would be to consider the heat capacities as constant, so that l becomes a linear function of T. With this approximation there may be obtained:

$$\log p = -\frac{2937.4}{T} - 4.9283 \log T + 23.5470 \tag{53}$$

(p in mb). This is sometimes called Magnus' formula; it corresponds to taking two terms within the bracket in Equation (51).

Table IV-1 compares some experimental values with the formulas (52) and (53). 0 °C has been taken as reference temperature, to fix the integration constant.

TABLE IV-1
Vapor pressure of water

t(°C)		−10	0	10	20	30
p from (52)	(mb)	2.875	6.11	12.32	23.7	43.6
p from (53)	(mb)	2.865	6.11	12.27	23.35	42.37
p observed	(mb)	2.863	6.11	12.27	23.37	42.43

For sublimation, the first approximation gives

$$\log p = 10.550 - \frac{2667}{T} \tag{54}$$

(p in mb). The second approximation might be obtained by using a constant value for the specific heat capacity of ice:

$$c_i = 2060 \text{ J kg}^{-1}\text{K}^{-1}$$

(as for water, no specification of constant pressure is necessary for ice). Due to the actual variation of c_i with temperature, however, it turns out that there is no advantage in using the second approximation*.

The table of physical constants at the end of the book contains a skeleton tabulation of thermodynamic properties of condensed water, in its solid and liquid phases.

4.9. Variation of Latent Heat Along the Equilibrium Curve

In Chapter II, Section 6, we studied the influence of temperature on enthalpy changes associated with changes of phase (i.e., on the latent heats of change of state). The influence of pressure was mentioned to be small for sublimation and vaporization, but appreciable for fusion. We are now in a position to easily calculate this term, and shall do so for the pressure change necessary to keep the two-phase system in equilibrium when the temperature is altered; that is, we shall find the variation in latent heat *along*

---

* More exact formulas for the thermodynamic properties of water substance can be found in the World Meteorological Organization tables, the Smithsonian Meteorological Tables and other references given in the Bibliography.

the equilibrium curve. The thermodynamic equation of state III-(42), applied to a change of phase, gives (taking the difference between the two phases):

$$\left(\frac{\partial \Delta H}{\partial p}\right)_T = \Delta V - T\left(\frac{\partial \Delta V}{\partial T}\right)_p = \Delta V - T\Delta(V\alpha_p) \tag{55}$$

where α_p is the isobaric coefficient of expansion and Δ refers to the difference of values between the two phases. We take (46) in the form

$$\frac{dp}{dT} = \frac{\Delta H}{T \Delta V} \tag{56}$$

so that

$$\left(\frac{\partial \Delta H}{\partial p}\right)_T \frac{dp}{dT} = \frac{\Delta H}{T}\left[1 - \frac{T}{\Delta V}\left(\frac{\partial \Delta V}{\partial T}\right)_p\right] =$$

$$= \frac{\Delta H}{T}\left[1 - \frac{T\Delta(V\alpha_p)}{\Delta V}\right] \tag{57}$$

and (cf. II-45):

$$\frac{d\Delta H}{dT} = \left(\frac{\partial \Delta H}{\partial T}\right)_p + \left(\frac{\partial \Delta H}{\partial p}\right)_T \frac{dp}{dT} =$$

$$= \Delta C_p + \frac{\Delta H}{T}\left[1 - \frac{T}{\Delta V}\left(\frac{\partial \Delta V}{\partial T}\right)_p\right] = \Delta C_p + \frac{\Delta H}{T}\left[1 - \frac{T\Delta(V\alpha_p)}{\Delta V}\right] \tag{58}$$

For vaporization ($\Delta H = L_v$) and sublimation ($\Delta H = L_s$) we can make the approximation

$$\Delta V \cong V_v = \frac{R^*T}{p} \tag{59}$$

which makes the last term of (58) zero. In the case of fusion ($\Delta H = L_f$) this term is generally not negligible.

4.10. Water Vapor and Moist Air

We shall now consider more carefully the type of gaseous phase with which we shall be concerned: a mixture of dry air and water vapor. In what follows we shall represent the water vapor pressure by e, and the partial pressure of dry air by p_d, leaving the symbol p for the total pressure. Subscripts w and i on the water vapor pressure will indicate saturation values with respect to liquid water and ice, respectively; subscript c will stand for any of the two condensed phases ($e_c = e_w$ or e_i, as the case may be).

Let us first consider pure water vapor. We have seen in Chapter I, Section 12 that dry air can be treated with good approximation as an ideal gas. The same is true for water vapor in the range of temperatures and partial pressures of meteorological interest. For each temperature water vapor will depart most from ideal behavior when it approaches saturation. Table IV-2 shows these maximal departures for several temperatures.

TABLE IV-2

$t(°C)$	$e_w V/R^* T$
−50	1.0000
0	0.9995
25	0.9980
50	0.9961

As these data indicate, the departure from ideal behavior reaches only a few tenths per cent at the most. Thus in virtually all cases we can assume, with fairly good approximation:

$$ev_v = R_v T \tag{60}$$

where v_v is the specific volume and $R_v = R^*/M_v$ is the specific constant for water vapor: $M_v = 18.015$ is the molecular weight of water. This gives for the gas constant:

$$R_v = 461.5 \text{ J kg}^{-1} \text{ K}^{-1}.$$

It is customary to express R_v as a function of the dry air constant R_d. As $R^* = M_d R_d = M_v R_v$ (cf. Chapter I, Section 11),

$$R_v = \frac{M_d}{M_v} R_d = \frac{1}{\varepsilon} R_d; \qquad \varepsilon = 0.622 \cong \frac{5}{8}. \tag{61}$$

Equation (60) becomes with this notation:

$$ev_v = \frac{1}{\varepsilon} R_d T. \tag{62}$$

For accurate computations, however, the departure from ideal conditions should be taken into account. Besides, we are interested in moist air, rather than pure water vapor. Two further effects occur with the addition of dry air, which modify appreciably the values of the saturation vapor pressures over water and ice. In the first place, we have to consider the displacement of the equilibrium caused by the increase in the total pressure. This can be computed by using Chapter III, Equation (25), which, written for unit mass, is

$$dg = -s \, dT + v \, dp. \tag{63}$$

The increase in specific Gibbs function of the condensed phase due to the increase in pressure at constant temperature will be:

$$\Delta g_c = \int_{e_c}^{p=e_c+p_d} v_c \, dp \cong v_c \, \Delta p \tag{64}$$

where $\Delta p = p_d \cong p$. Similarly, for the water vapor:

$$\Delta g_v = \int_{e_c}^{e_c + \Delta e_c} v_v \, de \cong v_v \Delta e_c \tag{65}$$

where the approximation can be made because Δe_c turns out to be very small.

Preservation of equilibrium implies that the chemical potentials of water in condensed phase and vapor must remain equal to each other. With adequate approximation this means equality in the Gibbs function (cf. Equation (10)). Therefore

$$\Delta g_c = \Delta g_v \tag{66}$$

and introducing Equations (64) and (65):

$$\Delta e_c = \frac{v_c}{v_v} \Delta p = \frac{v_c}{v_v} p_d. \tag{67}$$

For instance, at $0\,°C$ and $p_d \cong p = 1$ atm, $\Delta e_w = 0.005$ mb.

The second effect that air has on the water vapor equilibrium with the condensed phase arises from the small, but not entirely negligible, solubility of the gas in water; the solubility in ice is smaller and may be neglected. According to Raoult's law, this produces a decrease of the vapor pressure proportional to the molar fraction of dissolved gas.

Thus, we have three types of departure from the ideal case of pure water or ice in the presence of vapor behaving as an ideal gas: (1) total pressure is not the sum of the partial pressures of two ideal gases (Dalton's law of mixture of ideal gases: Chapter I, Section 11) as neither water vapor nor dry air are strictly ideal gases; (2) the condensed phase is under a total pressure augmented by the presence of the dry air; and (3) the condensed phase is not pure water substance, but contains dissolved gas. The three effects can be taken into account by an empirical correction factor f_c, which will be a function of both temperature and pressure. We can then write

$$e'_c(T, p) = f_c(T, p) \, e_c(T) \tag{68}$$

where $e_c(T)$ represents the vapor pressure of the pure condensed phase in absence of air, and is a function of the temperature alone, and $e'_c(T, p)$ is the corrected value

TABLE IV-3

Typical values of Empirical Correction factors

$t(°C)$	p(mb)	f_i 30	100	1100	$t(°C)$	p(mb)	f_w 30	100	1100
−80		1.0002	1.0008	1.0089	−40		1.0002	1.0006	1.0060
−40		1.0002	1.0006	1.0061	0		1.0005	1.0008	1.0047
0		1.0005	1.0008	1.0048	40			1.0019	1.0054

when air is present. Table IV-3 gives some values of f_i and f_w. We see from this table that even by assuming f_i and f_w to be unity we can ensure that the error is always less than 1%.

We can illustrate the effects of the presence of air on the saturation vapor pressures (second and third effects mentioned above) by considering how the triple point becomes modified. This is shown in Figure IV-5. P_t is the triple point of pure water

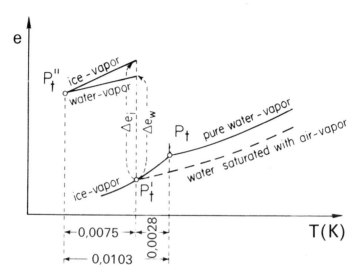

Fig. IV-5. Influence of air solubility and pressure on the triple-point equilibrium.

substance characterized by the temperature and pressure mentioned in Section 6. The effect of dissolved air is to lower the whole curve of water-vapor equilibrium; in the figure, the full curve becomes the lower dashed curve. The solubility of air gases in ice is truly negligible; therefore, if solubility was the only effect, the equilibrium ice-water-vapor would be now at the intersection of the dashed curve with the ice-vapor equilibrium curve, i.e. in P'_t. However, we must also consider the effect of pressure which will be different for the two condensed phases; according to Equation (66):

$$\Delta e_i = \frac{v_i}{v_v} p_d$$

$$\Delta e_w = \frac{v_w}{v_v} p_d.$$

These two increments must be added to the pressure corresponding to P'_t in order to obtain the new saturated vapor pressures over ice and over water; the former will be higher than the latter, because $v_i > v_w$ and therefore $\Delta e_i > \Delta e_w$. Through each of the two new points obtained, the corresponding equilibrium curves (ice-vapor and water-vapor, respectively) must pass; their slopes must be given by the two Clausius-Clapeyron equations (for ice and for water). With these slopes we extend the two curves until they intersect, at point P_t''. This is the new triple point, where ice, water saturated with air and water vapor will be in equilibrium, under a total pressure of $p = p_d + e_t''$ (the second term being the vapor pressure at P_t''). The changes in temperature indicated in the figure correspond to a total pressure of one atmosphere ($\sim 10^5$ Pa); solubility displaces the triple point by ~ 0.0028 K, and pressure by 0.0075 K, the total effect thus being a decrease in temperature of 0.010 K.

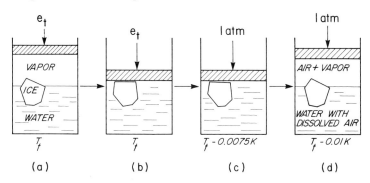

Fig. IV-6. Influence of air solubility and pressure on the triple-point equilibrium.

These relations may become clearer by performing an equivalent calculation with the help of an imaginary experiment. Figure IV-6 indicates schematically that in (a) we have a triple point equilibrium, with water substance only; e_t and T_t are the corresponding pressure and temperature. Now we condense all the vapor quasi-statically, at constant e_t and T_t (b). The pressure is then increased to $p = 1$ atm, while keeping equilibrium conditions; the system slides upwards on the fusion curve of Figure IV-1 (notice that the vertical axis variable is here the total pressure p, rather than the vapor pressure e, as in Figure IV-5). According to (47), the variation in temperature corresponding to the increase in pressure must be -0.0075 K; the system is in the situation (c). Finally, the piston is lifted while letting dry air come in, so as to preserve a total pressure of 1 atm (d). The liquid is under the same pressure as in (c), but air dissolves in the water, causing a new reduction in temperature of ~ -0.0028 K. The final temperature is $T_t - 0.01 = 273.15$ K. It will be noticed that (d) corresponds to the zero point chosen in the original definition of the Celsius scale (Chapter I, Section 5).

4.11. Humidity Variables

The water vapor content of moist air can be expressed through a number of different variables, several of which are of common use in meteorology. We shall first define the *specific humidity* q as the ratio of the mass of water vapor m_v to the total mass m:

$$q = m_v/m. \tag{69}$$

The specific gas constant for moist air will be:

$$R = \frac{m_d R_d + m_v R_v}{m} = (1-q)R_d + qR_v$$

$$= \left[1 + q\left(\frac{1}{\varepsilon} - 1\right)\right] R_d = (1 + 0.608\, q)R_d. \tag{70}$$

The equation of state becomes:

$$pv = (1 + 0.61\, q)R_d T = R_d T_v \tag{71}$$

by which we have defined the *virtual temperature* T_v

$$T_v = (1 + 0.61\, q)T \tag{72}$$

as the temperature of dry air having the same values of p and v as the moist air considered.

Another humidity variable is often used: the *mixing ratio r*, defined by

$$r = \frac{m_v}{m_d}. \tag{73}$$

Let us derive the relations between these two variables. From $m = m_d + m_v$ and the definitions of q and r, we obtain immediately

$$q = \frac{r}{1+r}; \quad r = \frac{q}{1-q}. \tag{74}$$

Both r and q are always smaller than 0.04. We may thus write, without great error,

$$q \cong r. \tag{75}$$

If we write both partial pressure equations for a given mass of moist air, and divide them side by side, we obtain:

$$p_d V = m_d R_d T$$

$$eV = \frac{1}{\varepsilon} m_v R_d T$$

$$\frac{e}{p_d} = \frac{r}{\varepsilon}.$$

That is:

$$r = \frac{\varepsilon e}{p - e}; \quad e = \frac{pr}{\varepsilon + r} \tag{76}$$

or approximately

$$r \cong \frac{\varepsilon}{p} e. \tag{77}$$

From $e = N_v p$ (cf. Chapter 1, Equation (21)) where N_v is the molar fraction of water vapor, comparing with Equation (76) (or directly from the definition of N_v) we see that

$$N_v = \frac{r}{\varepsilon + r}. \tag{78}$$

If the air is saturated with respect to water,

$$r_w = \frac{\varepsilon e_w}{p - e_w}; \qquad e_w = \frac{p r_w}{\varepsilon + r_w} \tag{79}$$

and with respect to ice:

$$r_i = \frac{\varepsilon e_i}{p - e_i}; \qquad e_i = \frac{p r_i}{\varepsilon + r_i}. \tag{80}$$

In the atmosphere, almost invariably

$$e_w < 60 \text{ mb}; \qquad r_w < 0.04$$

so that we may write as an approximation

$$e \cong \frac{p}{\varepsilon} r \tag{81}$$

$$e_w \cong \frac{p}{\varepsilon} r_w; \qquad e_i \cong \frac{p}{\varepsilon} r_i. \tag{82}$$

We shall also define another humidity variable, the *relative humidity*, as the ratio of the molar fraction of water vapor to the molar fraction corresponding to saturation with respect to water N_{vw} or with respect to ice N_{vi}. The corresponding relative humidities will be represented by U_w and U_i, respectively. Because we consider vapor and air as ideal gases, we can write

$$U_w = \frac{N_v}{N_{vw}} = \frac{p N_v}{p N_{vw}} = \frac{e}{e_w} \tag{83}$$

and similarly

$$U_i = \frac{e}{e_i}. \tag{84}$$

Within the usual ideal gas approximation, (83) and (84) can be taken as the definition of relative humidity.

To derive r as a function of U_w and r_w, we take the expression of r given by Equation (76), and we replace $e = U_w e_w$, and e_w by the second Equation (79). We obtain

$$r = \frac{U_w r_w}{1 + \frac{(1 - U_w) r_w}{\varepsilon}} \cong \tag{85}$$

$$\cong U_w r_w. \tag{86}$$

And similarly for U_i and r_i.

Usually U_w is obtained experimentally. p and T are also known. r_w is obtained from T through a table or a graph. q and r may then be calculated from the formulas above mentioned.

U_w and U_i are commonly given in percentage, q and r in units per mille (as grammes per kilogramme).

It should be noted that it is customary, for meteorological purposes, to express relative humidity at temperatures less than 0 °C with respect to water. The advantages of this procedure are as follows:

(a) Most hygrometers which are essentially responsive to the relative humidity indicate relative humidity with respect to water at all temperatures.
(b) The majority of clouds (between 0 °C and at least -20 °C) consist, in whole or in part, of supercooled water. Only below -40 °C are clouds always entirely glaciated.
(c) As a corollary to the above, supersaturation with respect to ice can frequently occur; this is not true with respect to water, so that it is convenient to require only two digits in coded messages for relative humidity.

4.12. Heat Capacities of Moist Air

In Chapter II, Section 5 we have seen the values for the heat capacities and ratios $\varkappa$ and η for diatomic gases, which may be applied to dry air.

The water vapor molecule is a triatomic non-linear molecule, whose position may be described by 3 translational and 3 rotational coordinates, giving 6 quadratic terms in the expression of its kinetic energy. Correspondingly, the equipartition theorem would give

$$c_{v_v} = \tfrac{6}{2} R_v = 3 R_v = 0.331 \text{ cal g}^{-1} \text{ K}^{-1}$$

and

$$c_{p_v} = 4 R_v = 0.441 \text{ cal g}^{-1} \text{ K}^{-1}$$

on the assumption that the vibrational energy does not contribute to the specific heat capacity (all molecules in ground state). However, spectra of atmospheric radiation show vibration-rotation bands produced by water vapor at atmospheric temperatures (particularly a strong vibration band centered at 6.27 μm). Larger experimental

TABLE IV-5

Specific heat capacities of ice (c_i), water (c_w) and water vapor at constant pressure (c_{p_v}) (IT cal g^{-1} K^{-1}).

t(°C)	c_i	c_w	c_{p_v} (pressure = e_i)	c_{p_v} (pressure = e_w)
−60	0.397	–	–	–
−40	0.433	1.14	0.4429	0.4430
−30	0.450	1.08	0.4434	0.4436
−20	0.468	1.04	0.4441	0.4443
−10	0.485	1.02	0.4451	0.4452
0	0.503	1.0074	0.4465	0.4465
10	–	1.0013	–	0.4482
20	–	0.9988	–	0.4503
30	–	0.9980	–	0 4530
40	–	0.9980	–	0.4552

values are therefore obtained, somewhat dependent on temperature and pressure. Table IV-5 gives some values for c_p, as well as for c_w and c_i.

If we take the specific heat capacities c_{v_v} and c_{p_v}, as well as those of dry air, as approximately independent of temperature, we may write to a good approximation

$$c_{p_v} = 0.447 \text{ cal g}^{-1}\text{K}^{-1} = 1870 \text{ J kg}^{-1}\text{K}^{-1}$$

$$c_{v_v} = 0.337 \text{ cal g}^{-1}\text{K}^{-1} = 1410 \text{ J kg}^{-1}\text{K}^{-1}.$$

If we now consider unit mass of moist air, the heat δQ absorbed at constant pressure for an increase dT in the temperature will be:

$$\delta Q = m_d \delta q_d + m_v \delta q_v = (1-q)\delta q_d + q\delta q_v$$

where δq_d and δq_v are the heats absorbed by unit mass of dry air and of water vapor, respectively, and q is the specific humidity. Dividing by dT:

$$c_p = (1-q)c_{p_d} + qc_{p_v} = c_{p_d}\left[1 + \left(\frac{c_{p_v}}{c_{p_d}} - 1\right)q\right]$$

$$= c_{p_d}(1 + 0.87q) \cong c_{p_d}(1 + 0.87r). \qquad (87)$$

Similarly we can obtain

$$c_v = c_{v_d}(1 + 0.97q) \cong c_{v_d}(1 + 0.97r) \qquad (88)$$

$$\varkappa = \frac{R}{c_p} = \frac{(1 + 0.61q)R_d}{(1 + 0.87q)c_{p_d}} = \frac{1 + 0.61q}{1 + 0.87q}\varkappa_d$$

$$\cong \varkappa_d(1 - 0.26q) \cong \varkappa_d(1 - 0.26r) \qquad (89)$$

$$\eta \cong \eta_d(1 - 0.10\,q) \cong \eta_d(1 - 0.10\,r) \tag{90}$$

where account has been taken of the fact that $q \ll 1$.

4.13. Moist Air Adiabats

According to Poissons's equations (Chapter II, Equation (53)), for an adiabatic expansion or compression from p, T to p', T', we must have

$$T' = T\left(\frac{p'}{p}\right)^{\varkappa}. \tag{91}$$

As $\varkappa$ varies with q for moist air, the adiabats passing through a point p, T will be different for different values of q. In particular $\varkappa = \varkappa_d$ for dry air, and as $\varkappa \leqslant \varkappa_d$, the adiabats in a diagram T, p will be slightly less steep (T will vary slightly more slowly) for moist air than for dry air.

The potential temperature of unsaturated moist air, θ_m, will be from Chapter II, Equation (55), and (89),

$$\theta_m \cong T\left(\frac{1000}{p}\right)^{\varkappa_d(1-0.26q)} \cong \theta\left(\frac{1000}{p}\right)^{-0.07q} \tag{92}$$

(p in mb). Differentiation yields

$$\frac{\partial \theta_m}{\partial q} \cong -0.07\,\theta \ln\left(\frac{1000}{p}\right). \tag{93}$$

Calculation with Equation (93) shows that the difference $(\theta_m - \theta)$ generally is less than $0.1\,°C$, so that one can treat unsaturated ascent or descent of air as if it were dry.

If in the definition of potential temperature

$$\theta = T\left(\frac{1000}{p}\right)^{\varkappa}$$

we put $\varkappa \cong \varkappa_d$ and substitute the virtual temperature T_v for T, the new expression defines the *virtual potential temperature* θ_v.

4.14. Enthalpy, Internal Energy and Entropy of Moist Air and of a Cloud

When we consider moist air as a closed system, and within the accepted approximations of ideal behavior, the values of the internal energy, enthalpy or entropy will be given by the expressions derived in Chapter II, Equations (37) and (38) and Chapter III, Equation (52). Referring to unit mass:

$$u = c_v T + \text{const.} \tag{94}$$

$$h = c_p T + \text{const.} \tag{95}$$

WATER-AIR SYSTEMS

$$s = c_p \ln T - R \ln p + \text{const.} \tag{96}$$

where R, c_p and c_v are given by Equations (70), (87) and (88).

But we are also interested in the study of clouds of water droplets or of ice crystals. We shall treat them as closed heterogeneous systems; each phase, as seen in Section 1, is an open subsystem. We may then apply Equation (16) to the enthalpy. It will be, assuming that the condensed phase is water:

$$\begin{aligned} dH &= \left(\frac{\partial H}{\partial T}\right)_{p,m} dT + \left(\frac{\partial H}{\partial p}\right)_{T,m} dp + (h_v - h_w) dm_v \\ &= \left(\frac{\partial H}{\partial T}\right)_{p,m} dT + \left(\frac{\partial H}{\partial p}\right)_{T,m} dp + l_v dm_v. \end{aligned} \tag{97}$$

The enthalpy of the system will be the sum of the enthalpies of the two components in the two phases:

$$\begin{aligned} H &= m_d h_d + m_v h_v + m_w h_w \\ &= m_d h_d + m_v (h_v - h_w) + m_t h_w \\ &= m_d h_d + m_v l_v + m_t h_w. \end{aligned} \tag{98}$$

Here the partial specific enthalpies $\bar{h}_k$ have been taken as the specific enthalpies h_k, as indicated in Section 1, and $m_t = m_v + m_w$ is the total mass of the water substance component. The conditions of a closed total system (implicit in Equation (97)) are expressed by

$$m_d = \text{const.}$$

$$m_t = m_v + m_w = \text{const.}$$

According to Equation (98), the partial derivatives in Equation (97) are really sums of the partial derivatives for the two components in the two phases, multiplied by the respective masses. Of these derivatives, we know that, within the ideal gas approximation,

$$\left(\frac{\partial h_d}{\partial T}\right)_{p,m} = c_{pd}, \quad \left(\frac{\partial h_v}{\partial T}\right)_{p,m} = c_{pv}, \quad \left(\frac{\partial h_d}{\partial p}\right)_{T,m} = \left(\frac{\partial h_v}{\partial p}\right)_{T,m} = 0.$$

For water,

$$\left(\frac{\partial h_w}{\partial T}\right)_{p,m} = c_w$$

where we do not need to specify "at constant pressure" in the specific heat, because c_p and c_v differ very little for water and ice (cf. Chapter III, Section 11). It may also be shown that

$$\left(\frac{\partial h_w}{\partial p}\right)_{T,m}$$

80   ATMOSPHERIC THERMODYNAMICS

which may be computed from Equation (42) in Chapter III, Section 8, is a negligible quantity*.

Introducing these values into Equation (97), we obtain

$$dH = (m_d c_{p_d} + m_v c_{p_v} + m_w c_w) dT + l_v dm_v$$
$$= mc_p dT + l_v dm_v \qquad (99)$$

where c_p is the mean specific heat, given by

$$mc_p = m_d c_{p_d} + m_v c_{p_v} + m_w c_w. \qquad (100)$$

If we now want to integrate the expression (99), to find the enthalpy of any given state, we must choose a reference state by specifying a certain value of the temperature and of the mass of water vapor m_v; the pressure does not need to be specified, because the expression (99) does not depend on it. The integration constant is then fixed by assigning the value $H = 0$ to the reference state. We shall choose a state at a temperature T_0 and with all the water in the liquid state.

We have to consider a process by which the system changes its temperature to T and a mass m_v of water goes into the gas phase. We may choose, among others, the two paths indicated below: a) first evaporating m_v grammes of water at a constant temperature T_0 and pressure, and then heating the whole system to T, or b) first heating the dry air and the water to T and then evaporating m_v grammes of water at T.

Temperature:	T_0		T_0		T
Mass of dry air:	m_d	a	m_d		m_d
Mass of water vapor:	0		m_v		m_v
Mass of liquid water:	m_t	b	m_w		m_w
Temperature:			T		
Mass of dry air:			m_d		
Mass of water vapor:			0		
Mass of liquid water:			m_t		

If we now integrate Equation (99) along both paths (inserting in each step the appropriate values of the temperature, or of the masses of vapor and water, and considering the specific heats as independent of the temperature), we obtain:

---

* $\left(\dfrac{\partial h_w}{\partial p}\right)_{T,m} = v_w(1 - \alpha_p T) \cong v_w \quad (\alpha_p = -6 \times 10^{-5} \text{ K}^{-1})$

This will contribute in (97) a term $m_w v_w dp$, which may be compared with the contribution $m_w c_w dT$ from the temperature derivative. We see by comparing these two terms that a change of as much as 1 atm in the pressure is equivalent to a temperature change as small as 1/40 K.

(a) $$H = H_0 + (m_d c_{p_d} + m_v c_{p_v} + m_w c_w)(T - T_0) + l_v(T_0)m_v$$
$$= H_0 + mc_p(T - T_0) + l_v(T_0)m_v \quad (101)$$

(b) $$H = H_0 + (m_d c_{p_d} + m_t c_w)(T - T_0) + l_v(T)m_v \quad (102)$$

where H_0 is the value of H in the reference state. Both expressions, which are equivalent, are related by Kirchhoff's theorem Chapter II, Section 6 which may be written here (in integrated form):

$$l_v(T) - l_v(T_0) = (c_{p_v} - c_w)(T - T_0). \quad (103)$$

It should be noticed that the coefficient of $(T - T_0)$ in Equation (101) varies with m_v, and $l_v(T_0)$ is a constant, while in Equation (102) the coefficient is a constant, and $l_v(T)$ is a function of T (l_v decreases about 0.1% for each degree of increase in temperature). In Equation (102) the terms with T_0 and H_0 can thus be written as an integration constant:

$$H = (m_d c_{p_d} + m_t c_w)T + l_v(T)m_v + \text{const.} \quad (104)$$

If the formula is referred to the unit mass of dry air, i.e., if the system contains the unit mass of dry air, its enthalpy can be written

$$H_1 = (c_{p_d} + r_t c_w)T + l_v(T)r + \text{const.} \quad (105)$$

where $r_t = m_t/m_d$ and the subscript of H indicates that its value is referred to unit mass of dry air.

If the heat capacities of the vapor and the water are considered as small quantities, m_t is assumed negligible as compared with m_d and l_v is taken as approximately independent of the temperature, we may write the formulas:

$$H_1 = h \cong c_p T + l_v q + \text{const.}$$
$$\cong c_p T + l_v r + \text{const.} \quad (106)$$

where h is the enthalpy of the unit mass of the total system (including both components), as approximate expressions for either Equations (101) or (102). Here the expressions (87) for moist air can be written for c_p.

The advantage of having an integrated expression like Equation (101), (102) or (104) lies in that we can calculate the difference ΔH between any two states of a closed system by simple difference, the integration having been done once and for all. Differentiation of these formulas will of course give back Equation (99) or an equivalent expression.

If we want to derive an expression for the internal energy, we start with

$$dU = \left(\frac{\partial U}{\partial T}\right)_{p,m} dT + \left(\frac{\partial U}{\partial p}\right)_{T,m} dp + (u_v - u_w)dm_v \quad (107)$$

instead of Equation (97). We may then notice that the derivatives with respect to p vanish as before, that the derivatives with respect to T give specific heats at constant volume (instead of at constant pressure; the one for liquid water being practically the same), and that

$$u_w = h_w - pv_w$$

$$u_v = h_v - ev_v = h_v - R_vT.$$

Following then the same integration as before, we obtain through the path b

$$U = (m_d c_{v_d} + m_t c_w)T + [l_v(T) - R_vT]m_v + \text{const.} \tag{108}$$

which could also be derived from Equation (104), with the appropriate substitutions ($h_d = u_d + R_dT$; $h_v = u_v + R_vT$; $h_w \cong u_w$; $c_{pd} = c_{v_d} + R_d$); using (98) and a similar expression for U).

For a system with unit mass of dry air:

$$U_1 = (c_{v_d} + r_t c_w)T + [l_v(T) - R_vT]r + \text{const.} \tag{109}$$

and with simplifications similar to the ones applied to the enthalpy, Equation (108) becomes

$$U_1 \cong u \cong c_vT + (l_v - R_vT)q + \text{const.} = c_vT + (l_v - R_vT)r + \text{const.} \tag{110}$$

$$\cong c_vT + l_vq + \text{const.} \cong c_vT + l_vr + \text{const.} \tag{111}$$

where U_1 and u have similar meanings to those of H_1 and h. In (111) the term R_vT has been neglected against the latent heat, considering that it only amounts to about 5 or 6% of l_v.

A similar derivation can be performed for the entropy, but now we can no longer disregard the effect of pressure, and the reference state must be specified with both the temperature and the pressure. We shall define it as two phases, isolated from each other, one consisting of m_d grams of dry air at temperature T_0 and pressure p_0, and the other consisting of m_t grams of liquid water at the same temperature (and pressure, although this parameter is immaterial for the condensed phase).

We shall now apply the total differential expression of the entropy separately for the two components in the two phases:

$$dS_i = \left(\frac{\partial S_i}{\partial T}\right)_{p,m} dT + \left(\frac{\partial S_i}{\partial p}\right)_{T,m} dp + s_i \, dm_i \tag{112}$$

i stands for d, v or w, and $S_i = m_i s_i$. We may notice that p is here the pressure of the particular component for the two gases, p_d and e, while for water it is the total pressure p exerted upon it. As before,

$$S = S_d + S_v + S_w = m_d s_d + m_v s_v + m_w s_w. \tag{113}$$

In applying Equation (112) to the dry air we notice that $dm_d = 0$, and that (cf. Chapter III, Section 7)

$$\left(\frac{\partial s_d}{\partial T}\right)_{p,m} = \frac{c_{pd}}{T}; \quad \left(\frac{\partial s_d}{\partial p}\right)_{T,m} = -\left(\frac{\partial v}{\partial T}\right)_{p,m} = -\frac{R_d}{p_d}.$$

Therefore:
$$dS_d = m_d ds_d = m_d c_{p_d} \, d \ln T - m_d R_d \, d \ln p_d. \tag{114}$$

We also notice that
$$\left(\frac{\partial s_v}{\partial T}\right)_{p,m} = \frac{c_{p_v}}{T}; \qquad \left(\frac{\partial s_w}{\partial T}\right)_{p,m} = \frac{c_w}{T}$$

$$\left(\frac{\partial s_v}{\partial p}\right)_{T,m} = -\frac{R_v}{e}; \qquad \left(\frac{\partial s_w}{\partial p}\right)_{T,m} = -\left(\frac{\partial v_w}{\partial T}\right)_{p,m} \cong 0,$$

so that
$$dS_v = m_v c_{p_v} \, d \ln T - m_v R_v \, d \ln e + s_v \, dm_v \tag{115}$$
$$dS_w = m_w c_w \, d \ln T + s_w \, dm_w. \tag{116}$$

We now make the sum $dS = dS_d + dS_v + dS_w$, taking into account that $dm_w = -dm_v$ and that
$$s_v - s_w = \frac{l_v}{T}.$$

We obtain
$$dS = (m_d c_{p_d} + m_v c_{p_v} + m_w c_w) \, d \ln T - m_d R_d \, d \ln p_d -$$
$$- m_v R_v \, d \ln e + \frac{l_v}{T} \, dm_v. \tag{117}$$

In order to integrate this expression, we consider now that the air undergoes the reversible processes
$$T_0, p_0 \to T, p_0 \to T, p_d$$
and the water substance the reversible processes

$$\begin{cases} \text{mass of vapor:} & 0 \\ \text{mass of water:} & m_t \\ \text{temperature:} & T_0 \\ \text{pressure:} & p_0 \end{cases} \to \begin{cases} 0 \\ m_t \\ T \\ p_0 \end{cases} \to \begin{cases} 0 \\ m_t \\ T \\ e_w(T) \end{cases} \to \begin{cases} m_v \\ m_w \\ T \\ e_w(T). \end{cases}$$

We integrate Equation (117) over the total change, and obtain
$$S = S_0 + (m_d c_{p_d} + m_t c_w) \ln \frac{T}{T_0} - m_d R_d \ln \frac{p_d}{p_0} + \frac{l_v(T) m_v}{T} \tag{118}$$

where S_0 is the entropy at the reference state*.

We have now two separate subsystems, one consisting of dry air at T, p_d and the other consisting of water in presence of its saturated vapor. The two gaseous phases occupy the same volume v, by hypothesis (within the ideal behavior assumption), and they can be mixed by an ideal process using semipermeable membranes, as described in Chapter III, Section 10, without change in entropy.

Provision should also be made in this ideal process, to perform the increase in pressure over the water from e_w to $p = p_d + e_w$, but this can be disregarded because the effect on the entropy is negligible.

We have thus finally the heterogeneous system in which we are interested: m_w grammes of liquid water in presence of a mixture of dry air and saturated vapor at the partial pressures p_d and e_w. Its entropy (referred to the reference state as defined) is given by Equation (118).

As we did for the enthalpy (Equation (104)), we can consider the terms with T_0, p_0 and s_0 as an undetermined integration constant, and write

$$S = (m_d c_{p_d} + m_t c_w) \ln T - m_d R_d \ln p_d + \frac{l_v(T) m_v}{T} + \text{const.} \tag{119}$$

The entropy S_1 of a system with unit mass of dry air can be written

$$S_1 = (c_{p_d} + r_t c_w) \ln T - R_d \ln p_d + \frac{l_v(T) r}{T} + \text{const.} \tag{120}$$

Making the same simplifications as before (Equation (106)) and considering that $p_d \cong p$, we obtain the approximate expressions

$$S_1 \cong s \cong c_p \ln T - R_d \ln p + \frac{l_v q}{T} + \text{const.} =$$

$$= c_p \ln T - R_d \ln p + \frac{l_v r}{T} + \text{const.} \tag{121}$$

where s is the entropy of unit mass of cloud.

We have derived the expression for a system with the vapor pressure saturated at the given temperature, because it is the case in which we shall be interested. By separating the water phase and adding an appropriate expansion of the vapor in the series of steps imagined for the water substance in the derivation, it could be easily seen that a term $-m_v R_v \ln(e/e_w)$ should be added to Equation (119) if the vapor pressure of the system is e rather than e_w; obviously, such a system could not be in equilibrium.

---

* We may notice that by differentiation of this expression, one obtains

$$ds = (m_d c_{p_d} + m_t c_w) \, d\ln T - m_d R_d \, d\ln p_d + \frac{l_v(T)}{T} dm_v - \frac{l_v(T) m_v}{T^2} dT + \frac{m_v}{T} dl_v(T).$$

This is an alternative formula for Equation (117) and can be seen to coincide with it by taking into account the Clausius-Clapeyron equation $d \ln e/dT = l_v(T)/R_v T^2$ and Kirchhoff's law $dl_v = (c_{p_v} - c_w) dT$.

PROBLEMS

1. The average value of the heat of vaporization of water in the temperature interval 90–100 °C is 542 cal g^{-1}. Derive the value of the water vapor pressure at 90 °C.

2. From the equation

$$\log_{10} e = -\frac{2937.4}{T} - 4.9283 \log_{10} T + 23.5471$$

   for the saturation vapor pressure of water (where e is given in mb), derive its latent heat of vaporization at 10 °C.

3. The specific volume of liquid water is 1.000 cm^3 g^{-1}, and that of ice, 1.091 cm^3 g^{-1}, at 0 °C. What is the rate of change of the melting point of ice with pressure in K atm^{-1}?

4. (a) The formula

$$\log_{10} e = 9.4041 - \frac{2354}{T}$$

   gives the water vapor pressure (e in mb) as a function of the temperature. Using this expression, whose constants have been adjusted to give the experimental value $e = 6.11$ mb for 0 °C, and the value of R^* from tables, derive the change in entropy when 1 mol of water evaporates in conditions close to equilibrium, at 0 °C. Give the result in cal K^{-1} mol^{-1}.
   (b) How much should be added to ΔS if the process takes place at 25 °C?

   Use the value $e_{25\,°C} = 31.67$ mb. Assume that l_v is unknown.

5. The melting point of ice is depressed by 0.075 degrees when the pressure is increased by 10 atm. From this information derive the value of the latent heat of freezing. The density of ice is 0.917 g cm^{-3}.

6. An air mass has a temperature of 30 °C and a relative humidity of 50% at a pressure of 1000 mb. Derive the values of: water vapor pressure, mixing ratio, specific humidity, specific heat capacity at constant pressure, virtual temperature, and the coefficient $\varkappa$. What would be the values of its potential temperature and its virtual potential temperature after expanding the air adiabatically to 900 mb.

7. A mass of moist air is at a pressure of 900 mb and a temperature of 2.1 °C. The mixing ratio is $r = 3 \times 10^{-3}$. Compute T_v, R, c_p and $\varkappa$.

8. The saturation vapor pressure at 25 °C is 31.67 mb.
   (a) What is the partial vapor pressure of a parcel of air at that temperature, if its dew point is 5 °C? (The dew point is the temperature at which the air becomes saturated, when cooled isobarically). Use neither diagrams nor the vapor pressure table.
   (b) What is the value of the mixing ratio r, if the pressure is 1000 mb?
   (c) What will be the values of the vapor pressure and the mixing ratio, if the parcel expands to 800 mb?

9. Knowing that saturated air at 30 °C and 1000 mb has a mixing ratio $r_w = 27.7$ g kg^{-1}, and that the average value of the latent heat of vaporization between 30 °C and 0 °C is 588.4 cal g^{-1}, calculate the saturation mixing ratio at this last temperature and 500 mb.

10. Using the value of the latent heat of fusion from tables, calculate the ratio e_w/e_i at -10 °C (e_w = saturation vapor pressure with respect to water; e_i = id. with respect to ice).

11. 1 g of dry air is saturated with water vapor at 20 °C.
    (1) What is the specific change in enthalpy and in entropy of the moist air, when cooled at 1 atm of total pressure from 20 °C to 0 °C, condensation taking place?
    (2) What are the relative errors in Δh and Δs made in the following approximations:
        (a) Assuming l_v = const. = 590 cal g^{-1}:
        (b) Neglecting the heat capacity of water?
        (c) Taking $m = m_d$? (m: total mass; m_d: mass of dry air).

12. Derive Gibbs' phase rule (Equation (38)).

CHAPTER V

EQUILIBRIUM WITH SMALL DROPLETS AND CRYSTALS

In this chapter we shall deal with a particular type of heterogeneous system, which is important in cloud physics. The reader who is not interested in that field can omit the chapter without loss of continuity.

We shall consider the equilibria associated with changes of phase in systems with one or two components and two phases, as those considered in the previous chapter, but with the difference that one of the phases is in such a state of subdivision that the effects of interfacial tensions cannot be ignored. We shall be interested, in particular, in obtaining expressions for the vapor pressure of small droplets and crystals, and for the temperature at which small crystals are in equilibrium with the liquid phase.

The general difference with the systems studied so far lies in that the work δA performed by external forces on the system consists not only of the work of expansion $-p\,dV$ but also of another term giving the work necessary to increase the area of the interface separating the condensed from the gaseous phase. By definition of interfacial tension, this term must be written $\sigma\,d\mathscr{S}$ where σ is the interfacial tension and $\mathscr{S}$ the area of the interface. The differential of the Gibbs function in a reversible process becomes in this case

$$dG = -S\,dT + V\,dp + \sigma\,d\mathscr{S} \tag{1}$$

5.1. Vapor Pressure of Small Droplets of a Pure Substance

In Chapter IV, Section 6, the vapor pressure of a liquid was defined as that pressure at which the vapor is in equilibrium with the liquid phase. It was implicitly understood in that definition that the separation surface between the phases should be plane. If the surface has a pronounced curvature, the vapor pressure depends on it; the pressure will be higher for convex surfaces, and smaller for concave surfaces, than for a plane one. The reason for this, as we shall see, is that the evaporation from a convex surface decreases its area, which contributes negatively to the variation of Gibbs function, while the opposite occurs with a concave surface. The curvature of an element of surface is described, in general, by means of two radii of curvature; however, we shall restrict consideration to the case when both radii are equal, i.e., to spherical surfaces. We shall be particularly interested in the case of small droplets suspended in gaseous phase and in these conditions the drops adopt a spherical shape*.

---

* When the drops are bigger, the falling velocity becomes appreciable and the drop may become deformed. But when this occurs the effect of curvature on vapor pressure is already negligible.

Let us now consider a spherical drop of radius R, in equilibrium with the vapor, which will be at the pressure p_R. We shall also consider a plane liquid surface, in equilibrium with its vapor at the pressure p_s (where the subscript s indicates saturation with respect to the liquid with a plane surface).

Let us assume now that we transfer dn moles of liquid from the mass with a plane surface to the drop. We can imagine two paths (A and B) to perform this operation:

(A) (a) Evaporation of dn moles from the mass with a plane surface, at constant pressure p_s and temperature T;
  (b) Isothermal compression of the vapor from p_s to p_R;
  (c) Condensation of the dn moles on the drop at constant pressure p_R and temperature T;

(B) (a) dn moles are separated mechanically from the mass with plane surface;
  (b) the dn moles are incorporated into the drop.

We must calculate the variation of Gibbs function of the total system (drop and mass with plane surface) for the considered modification (transference of the dn moles) by the two described paths.

We observe that in the first path the steps (a) and (c) do not contribute to the change in Gibbs function, because they occur reversibly at constant pressure and temperature. For step (b) we assume ideal behaviour for the vapor, and calculate*:

$$dG_A = dn \int_{p_s}^{p_R} V \, dp = dn R^* T \ln \frac{p_R}{p_s} \qquad (2)$$

where V represents the molar volume.

In order to calculate the variation dG_B corresponding to the second path, we consider first the variation of surface and volume when a drop of radius R increases by dn moles. Let n be the total number of moles in the drop, M the molecular weight of the liquid, ϱ its density and $\mathscr{V}$ the volume of the drop. We shall have:

$$\mathscr{V} = \frac{nM}{\varrho} = \tfrac{4}{3}\pi R^3, \qquad \mathscr{S} = 4\pi R^2$$

where $\mathscr{S}$ is the surface of the drop. We obtain by differentiating:

$$d\mathscr{V} = \frac{M}{\varrho} dn = 4\pi R^2 \, dR, \qquad d\mathscr{S} = 8\pi R \, dR$$

---

* Notice that the term with surface tension, which does not act as an external force, does not contribute to dG. During the condensation on the drop, which is the step in which its surface increases, we can imagine both the drop and the surrounding vapor subject, as a whole, to a pressure p_R as the sole external force acting on the system.

and, comparing both expressions,

$$d\mathscr{S} = \frac{2M}{\varrho R} dn \tag{3}$$

The mass with a plane surface can be considered as a part of a drop of infinite radius. $d\mathscr{S}$ is in that case a second-order infinitesimal; for the small drop, formula (3) will be valid with its corresponding value of R. The only work performed on the system in the step (B.a) is the work necessary to separate the dn moles (thereby creating an interface) from the mass with a plane surface. This work must be compensated in step (B.b) (where that interface disappears again); but in this last step the system must also receive the necessary work to increase the surface by $d\mathscr{S}$. According to (1), this work appears as the only term in the variation of Gibbs function because the process occurs at constant pressure and temperature. Therefore:

$$dG_B = \sigma \, d\mathscr{S} = \frac{2M\sigma}{\varrho R} d\mathscr{S}. \tag{4}$$

Equation (4) expresses the fact that the only difference between the values of Gibbs function for the initial and the final system (which only differ from the first in having dn fewer moles in the mass with a plane surface and dn more moles in the drop) is that corresponding to the difference in the drop surface area. As the variation in Gibbs function must be equal for both paths, we equate Equations (2) and (4) and obtain:

$$\ln \frac{p_R}{p_s} = \frac{2M\sigma}{\varrho R^* T R} = \frac{P}{R} \tag{5}$$

where $P = 2M\sigma/\varrho R^* T$; i.e.,

$$p_R = p_s e^{P/R} \tag{6}$$

If R is not very small, Equation (6) can be approximated by developing the exponential to the first-order term:

$$p_R \cong p_s \left(1 + \frac{P}{R}\right) \tag{7}$$

For water at $0\,°C$ we have:

$$\sigma = 0.0757 \text{ N m}^{-1}$$
$$\varrho = 1000 \text{ kg m}^{-3}$$
$$M = 0.01802 \text{ kg mol}^{-1}$$
$$P = 1.20 \times 10^{-9} \text{ m} = 1.20 \times 10^{-3} \ \mu\text{m}$$

with which are calculated the values of Table V-1 for the ratio p_R/p_s.

We may remark that the range of R values corresponding to water droplets in clouds goes from 1 to 100 μm.

The theory assumes that the surface tension is not affected by the curvature. This hypothesis becomes doubtful for radii much smaller than 0.01 μm.

TABLE V-1
Vapor pressure of small water drops

R (μm)	p_R/p_s
100	1.000012
10	1.00012
1	1.0012
0.1	1.012
0.01	1.128
0.001	3.32

As a consequence of the variation in vapor pressure with the size of drops, a system consisting of vapor and drops of various sizes cannot be in equilibrium; the bigger drops must grow at the expense of the smaller ones. In other words, the liquid distills from the smaller to the bigger drops until the former disappear.

This theory is still valid in the presence of an inert gas (e.g., air in a system of water drops and water vapor) and is therefore applicable to clouds. We must only take into account, in that case, that the pressures considered are the partial pressures of water vapor.

We shall call *saturation ratio r*, the ratio of water vapor pressure to the saturation vapor pressure $r = p/p_s$, and *supersaturation*

$$\frac{p}{p_s} - 1 = r - 1.$$

When air is present, r becomes equal to the relative humidity U_w. For instance, we can see in Table V-1 that, for an atmosphere saturated with respect to water droplets to have a supersaturation of 1%, the droplets must not be larger than approximately 0.1 μm radius.

In Figure V-2, the upper curve represents the values of the equilibrium saturation ratio $r_R = p_R/p_s$, as a function of the radius R, for pure water.

5.2. Vapor Pressure of Solution Droplets

We shall consider now the case of small drops of solutions of a non-volatile solute in a solvent. This is the case, for instance, of a salt dissolved in water and therefore it finds application in the study of cloud physics, because the droplets of the cloud form by condensation of vapor on nuclei of hygroscopic substances suspended in the air.

We assume now that we have a mass of pure solvent with a plane surface and a spherical drop of solution. We shall call p'_R the vapor pressure of the drop and σ' its surface tension (primed symbols will be used for the solution and unprimed symbols for the pure solvent). We shall imagine, as before, that we transfer dn_1 moles of solvent from the mass with plane surface to the drop; dn_1 is infinitesimal and it therefore does not alter the composition of the drop. With the subscript 1 we characterize the solvent, while subscript 2 will be used for the solute.

The first term will be, as before, the isothermal evaporation at pressure p_s, the isothermal compression (or expansion) to p'_R and the isothermal condensation over the drop at pressure p'_R. The variation in Gibbs function (assuming ideal behaviour) will have the value

$$dG_A = dn_1 R^* T \ln \frac{p'_R}{p_s}. \tag{8}$$

Path B will consist now of the following steps:

(a) dn_1 moles of pure solvent are separated from the mass with a plane surface. To do this it is necessary to create a new interface with an increase of the Gibbs function. But, as in the previous section, this will be compensated by the disappearance of that interface in step (c).

(b) The pressure over the drop is modified isothermally from p'_R to $p_s + \Pi$, where Π is the osmotic pressure of the drop solution*.

(c) The dn_1 moles are introduced reversibly into the drop through a semipermeable membrane that only permits the pure solvent to pass through. In this step, the interface created in (a) disappears, and the surface area of the drop increases. Schematically, we can represent the process as in Figure V-1.

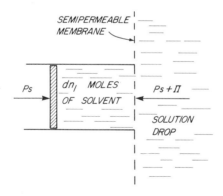

Fig. V-1. Reversible introduction of solvent into solution drop.

(d) The pressure over the drop is brought again, isothermally, to the value p'_R.

Let us calculate now the variation in Gibbs function by this path. The variation in step (a) is compensated by that in step (c). In step (b) we have

$$dG_{Bb} = \int_{p'_R}^{p_s + \Pi} V' \, dp$$

---

* By definition, Π is the difference between the pressure acting at both sides of a semi-permeable membrane which only permits the passage of pure solvent and that separates the solution from the pure solvent (i.e., the pressure over the solution less the pressure over the solvent), in equilibrium conditions.

where V' is the volume of the drop. In step (c), which occurs at constant pressure and temperature*, the drop surface increases, so that according to Equation (1):

$$dG_{Bc} = \sigma' \, d\mathscr{S}.$$

Finally, in step (d) we have

$$dG_{Bd} = \int_{p_s + \Pi}^{p'_R} (V' + dn_1 \bar{V}_1) \, dp$$

where $\bar{V}_1$ is the partial molar volume of the solvent in the solution (see Chapter IV, Section 1).

The sum of the three differentials gives

$$dG_B = dn_1 \int_{p_s + \Pi}^{p'_R} \bar{V}_1 \, dp + \sigma' \, d\mathscr{S}. \tag{9}$$

As a first approximation, we can take $\bar{V}_1$ as independent of the pressure and equal to the molar volume of the pure solvent V_1; the first term then becomes

$$dn_1 \int_{p_s + \Pi}^{p'_R} \bar{V}_1 \, dp \cong dn_1 V_1 (p'_R - p_s - \Pi) \cong -V_1 \Pi \, dn_1$$

where, in the last expression, we have considered that the difference $(p'_R - p_s)$ is negligible against Π. As to the second term, we consider that

$$V' = n_1 \bar{V}_1 + n_2 \bar{V}_2 = \tfrac{4}{3}\pi R^3$$

where $\bar{V}_2$ is the molar partial volume of the solute in the solution, and n_1 and n_2 the number of moles of solvent and solute in the drop (cf. Equation (8) of Chapter IV). As the solvent increases, the volume of the drop will increase by:

$$dV' = V_1 \, dn_1 = 4\pi R^2 \, dR.$$

Then:

$$d\mathscr{S} = 8\pi R \, dR = \frac{2\bar{V}_1 \, dn_1}{R} \cong \frac{2V_1 \, dn_1}{R}$$

and

$$\sigma' \, d\mathscr{S} = \frac{2V_1 \sigma' \, dn_1}{R}. \tag{10}$$

---

* We remark that the process can be assumed reversible, even though the pressures and the chemical potentials of the solute are different on both sides of the membrane, because the latter is neither deformable nor permeable to the solute.

Now we can equate the total variation of Gibbs function for the two paths considered (Equations (8) and (9)). It will be

$$dn_1 R^* T \ln \frac{p'_R}{p_s} = \frac{2V_1 \sigma' dn_1}{R} - \Pi V_1 dn_1$$

or

$$R^* T \ln \frac{p'_R}{p_s} = \frac{2V_1 \sigma'}{R} - \Pi V_1 \qquad (11)$$

We must substitute here an expression for Π. In order to derive it, it is sufficient to consider Equation (11) for the particular case when the solution is, like the solvent, in a large mass without appreciable curvature at its surface: $R = \infty$, $p'_R = p'_s$, and

$$\Pi V_1 = -R^* T \ln \frac{p'_s}{p_s}. \qquad (12)$$

According to Raoult's law, assuming that the solute is non-volatile,

$$\frac{p'_s}{p_s} = N_1 \qquad (13)$$

where N_1 is the molar fraction of the solvent in the solution.
Therefore:

$$\Pi V_1 = -R^* T \ln N_1 \qquad (14)$$

which is the thermodynamic expression usually derived for the osmotic pressure Π*.
Substituting in Equation (11), we obtain:

$$\ln \frac{p'_R}{p_s} = \frac{2\sigma' V_1}{R^* T R} + \ln N_1. \qquad (15)$$

And writing $V_1 = M/\varrho$, the equation becomes

$$\ln \frac{p'_R}{p_s} = \frac{2\sigma' M}{R^* T \varrho R} + \ln N_1$$

or else

$$\frac{p'_R}{p_s} = r_R = N_1 \exp(2\sigma' M / R^* T \varrho R). \qquad (16)$$

In this formula, we must take into account that, if the solute is an electrolyte, its concentration must be corrected by multiplying the number of moles n_2 by the van 't Hoff factor i. Therefore, if N_2 is the molar fraction of the solute:

$$N_2 = \frac{in_2}{n_1 + in_2}; \quad N_1 = \frac{n_1}{n_1 + in_2}.$$

---

* Other current expressions are derived from this one for dilute solutions, in which $n_2 \ll n_1$ and $\ln N_1 = \ln(1 - N_2) \cong -N_2$.

For dilute solutions of strong electrolytes, i is equal to the number of ions into which each molecule dissociates. Thus, for NaCl, dissociated into Cl^- and Na^+, $i = 2$. For concentrated solutions, i can vary appreciably from this value.

We can replace N_1 in Equation (16):

$$N_1 = 1 - N_2 = 1 - \frac{in_2}{n_1 + in_2}$$

and for dilute solutions, $in_2 \ll n_1$,

$$N_1 \cong 1 - \frac{in_2}{n_1} = 1 - \frac{im_2 M}{M_2 (4/3)\pi R^3 \varrho} = 1 - \frac{Q}{R^3} \qquad (17)$$

where m_2 is the mass of solute, M_2 its molecular weight, and

$$Q = \frac{3im_2 M}{4\pi \varrho M_2}.$$

If the drop is not very small, the exponent in Equation (16) is small and we can write:

$$\exp(2\sigma' M/R^* T\varrho R) \cong 1 + \frac{2\sigma' M}{R^* T\varrho R} = 1 + \frac{P}{R}$$

where

$$P = \frac{2\sigma' M}{R^* T\varrho}.$$

Thus, for not very small drops of not very concentrated solution,

$$r_R = \frac{p'_R}{p_s} \cong \left(1 + \frac{P}{R}\right)\left(1 - \frac{Q}{R^3}\right) \cong 1 + \frac{P}{R} - \frac{Q}{R^3} \qquad (18)$$

where we have assumed that the terms P/R and Q/R^3 are both small.

The coefficient P depends on temperature and (through σ') on the nature and concentration of the solution. Q is almost independent of temperature (ϱ is only slightly dependent), but depends on the nature and mass of the solute; therefore, Q will be constant for any droplet formed on a hygroscopic nucleus of a specified mass, which is the important case for the physics of clouds.

If the concentration does not exceed about 1%, we can assume $\sigma' \cong \sigma$, and P has the value given in Section 1. As for Q, it becomes, writing $i = 2$ (mono-monovalent, strong electrolyte):

$$Q = 8.6 \frac{m_2}{M_2} \text{ cm}^3.$$

If the solute is sodium chloride:

$$M_2 = 58.45 \text{ g mol}^{-1}$$

$$Q = 0.147 \, m_2 \text{ cm}^3$$

where m_2 is expressed in grams.

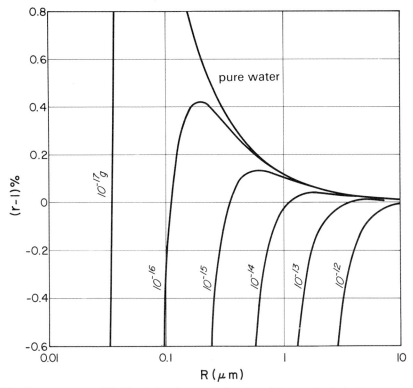

Fig. V-2. Vapor pressure of NaCl solution drops, as a function of drop radius R. $(r-1)$: supersaturation. Numbers on curves indicate mass of NaCl.

Figure V-2 represents the saturation vapor pressure as a function of the radius R for NaCl and $T = 273$ K; the different curves correspond to different values of the mass m_2. The upper, right-hand curve corresponds to pure water.

We can see in the figure that the curves pass through a maximum (except for pure water). Let us consider a drop of large radius R containing a certain mass of solute m_2. Its vapor pressure will be very close to p_s $(r \cong 1)$. If we now assume that the water gradually evaporates, R will decrease (we shift towards the left in the graph); the effect of curvature will become noticeable and, with it, the term P/R in Equation (18): p'_R will increase approximately following the pure water curve, because the term Q/R^3 is still very small. However, as R decreases, this last term increases as R^{-3}, while P/R increases only as R^{-1}. A stage will be reached, therefore, where the term Q/R^3 becomes first noticeable, then grows rapidly and soon becomes more important than P/R: the curve diverges from that of pure water, passes through a maximum and then decreases rapidly. The curve will finish at that value of the saturation ratio r corresponding to the vapor pressure of the saturated solution of NaCl ($r = 0.78$); in this last part (left branch of the curve) the approximation (18) ceases to be valid because the solution is too concentrated.

If the solution drop is in an atmosphere kept at a value of r lower than the maximum of the curve, it will become stable with the corresponding value of the radius R, as given by the curve. If r is higher than the maximum, the drop will start growing indefinitely, while its vapor pressure p'_R decreases along the right side of the curve. In this sense, the values corresponding to the maximum, r_c and R_c, are critical values for the activation of hygroscopic nuclei in the atmosphere, leading to the formation of cloud drops.

In order to obtain the critical values, we should differentiate Equation (16) with respect to R and equate to 0. But it is easy to verify that the maximum is already in a region in which the solution is dilute enough to apply the approximate Equation (18). Differentiating the latter and equating to 0, we obtain:

$$R_c = \sqrt{\frac{3Q}{P}} \qquad (19)$$

which, substituted into Equation (18), gives:

$$r_c = 1 + \frac{2\sqrt{P^3}}{3\sqrt{3Q}} \qquad (20)$$

Table V-2 gives the critical values for several sizes of sodium chloride nuclei. For instance, a nucleus of 10^{-15} g will become activated with a supersaturation of 0.13%, in a drop of radius 0.62 μm.

TABLE V-2
Critical activation values (drop radius and supersaturation) for sodium chloride nuclei

m_2 (g)	R_c (μm)	$r_c - 1$
10^{-16}	0.19	4.2×10^{-3}
10^{-15}	0.62	1.3×10^{-3}
10^{-14}	1.9	4.2×10^{-4}
10^{-13}	6.2	1.3×10^{-4}
10^{-12}	19.0	4.2×10^{-5}

5.3. Sublimation and Freezing of Small Crystals

We can make, step by step, the same derivation that we made for condensation in Section 1, for the phenomenon of sublimation. Therefore we can write:

$$\ln \frac{p_R}{p_s} = \frac{2\sigma_{sv} M}{\varrho_s R^* T R} \qquad (21)$$

where p_R and p_s are now referred to the solid, σ_{sv} is the surface tension of the solid in the presence of its vapor, and ϱ_s is the density of the solid. This equation implies the approximation of considering the small crystal as if it had spherical shape, because it has been assumed that

$$d\mathscr{S} = \frac{2}{R} dV.$$

If the relation between $d\mathscr{S}$ and dV is different from $2/R$, we should substitute this factor in the formula by another factor which will depend on the shape of the crystal.

For the melting equilibrium (small crystals in equilibrium with the liquid) a similar derivation which will not be included here shows that the equilibrium temperature T is lower than that corresponding to a macroscopic plane surface of the crystal (T_0), and the difference is given by the formula

$$T_0 - T = \frac{2\sigma_{sl} T_0}{l_f \varrho_s R} \qquad (22)$$

where σ_{sl} is the interfacial tension solid-liquid, l_f is the latent heat of fusion, and the shape of the crystal has been again approximated by a sphere of radius R. The difference $(T_0 - T)$ is very small, except for extremely small values of R; for example, for ice, it reaches 1 K when R becomes equal to 0.036 μm (accepting a value $\sigma_{sl} = 20$ dyn cm^{-1} for the interfacial tension).

PROBLEMS

1. Calculate the radii of water droplets in equilibrium with an atmosphere whose supersaturation is (a) 1%; (b) 0.1%; (c) 0.05%. Assume a temperature of 0 °C.
2. Consider a nucleus of NaCl of mass 3×10^{-14} g. Derive:
   (a) The radius of a droplet containing this nucleus in solution, for which the vapor pressure e'_r is exactly equal to that of pure water with a plane surface e_s.
   (b) The critical radius, over which the nucleus becomes activated. The two results will be in the order of magnitude of micrometers. You can use this fact to simplify the calculations. The temperature is 0 °C. The van't Hoff factor is $i = 2$.
3. A water droplet containing 3×10^{-16} grams of sodium chloride has a radius of 0.3 μm.
   (a) Calculate its vapor pressure.
   (b) If the droplet is in equilibrium with the environment, what is the supersaturation (expressed in percentage) of this environment?
   (Note: In computing molar ratios, each formula weight ('molecular weight' M_{NaCl}) of NaCl must be considered as 2 moles, because of the total dissociation in solution; i.e. the van't Hoff factor is $i = 2$). The temperature is 25°C. At that temperature, the saturation vapor pressure is $e_s = 31.67$ mb.

CHAPTER VI

AEROLOGICAL DIAGRAMS

6.1. Purpose of Aerological Diagrams and Selection of Coordinates

In order to be able to study with speed and convenience the vertical structure and a number of properties of the atmosphere above a certain location, use is made of special thermodynamic diagrams, or *aerological diagrams*. The observational data to be represented are obtained from soundings, and consist of sets of values of temperature, pressure, and humidity (e.g., relative humidity).

These data are considered as essentially vertical and instantaneous. With a net of sounding stations, they can be organized in a tridimensional description of the atmosphere, and with successive soundings, evolution with time can be followed.

The system to be represented on these diagrams will be in general moist air. It has two components and one phase; its variance, according to the phase rule, must be three, corresponding to the three different variables measured in the soundings. As there are three independent variables, p, T and U_w (or U_i) that have to be plotted on a plane surface, the representation should consist of one curve with a set of scalars (e.g., p as a function of T, and values of U_w written for points along the curve), or two different curves (e.g., T as a function of p and U_w as a function of p). If there is water (or ice) present, as in a cloud, the variance is reduced to two, and one curve is enough (e.g., T as a function of p; $e = e_w(T, p)$ is not independent).

On a thermodynamic diagram, different *isolines* or *isopleths* can be drawn: isobars, isotherms, equisaturated curves or 'vapor lines' (curves for which $U_w = 1$), curves of equal potential temperature or isentropics (cf. Chapter III, Section 9) for dry air, or 'dry adiabats', etc. The ones specifically mentioned and the 'saturated adiabats', to be studied later, are the most important, and in that sense we shall call them the *fundamental lines* of the diagram. It should be remarked that these lines, as any other curve drawn on the diagram, may have two meanings. One, static, indicating the vertical structure of an atmospheric layer – thus, an adiabat may represent an atmosphere with constant value of the potential temperature (and of the specific entropy) along the vertical. But it may also have the meaning of a *process curve*, representing the change of the variables for an air parcel undergoing an adiabatic expansion as it rises through the atmosphere. The coincidence of representation means that the potential temperature is an invariant for the process just mentioned.

The importance of the diagrams lies in the large amount of information that can be rapidly obtained from them. They allow the study of the vertical stability of the atmosphere. The thickness of layers between two given values of the pressure is readily computed from them. A number of atmospheric processes can be conveniently studied with their aid. The vertical structure will also indicate the type of air mass or air masses involved.

These uses determine the choice of the coordinates, and many different diagrams have been developed for general and for special purposes. We shall only see a few of the most commonly used. There are certain criteria which are taken into account in the selection of an appropriate diagram; we enumerate them below.

(a) The angle between isotherms and adiabats. The larger this angle is, the more sensitive is the diagram to variations in the rate of change of temperature with pressure along the vertical. This is important, as we shall see, for the analysis of stability, including synoptic aspects of frontal and air mass analysis.

(b) Which and how many isopleths are straight lines. Straight lines facilitate the use of the diagram for plotting as well as for analysis and representation.

(c) If energy integrals (such as, for instance, the work performed by a parcel of moist air along a cycle) can be determined by measuring areas on the diagram. Such diagrams are sometimes called *equivalent*, or *area-preserving*.

(d) Some advantage may be obtained if one of the main isopleths is congruent with respect to a displacement along one of the coordinates (cf. Sections 3 and 5).

(e) If the ordinate varies monotonically with height, roughly proportional to it, the atmosphere can be more conveniently imagined or visualized over the diagram.

6.2. Clapeyron Diagram

Clapeyron diagram uses the coordinates $-p, v^*$. It is the diagram of frequent use in general thermodynamics, with a changed sign for p, to comply with condition (e) of Section 1. As for the other conditions (cf., Figure VI-1):

(a) The angles between adiabats and isotherms are small.

(b) Isobars and isochores are straight lines. Adiabats and isotherms are curves.

(c) It is equivalent; in a cycle: $a = \oint (-p)\,dv$.

This is not a convenient diagram and therefore it is not used. We mention it here only for comparison purposes, as it is a well-known representation in thermodynamics.

6.3. Tephigram

The coordinates are $\ln \theta, T$. Its name, given by Shaw, comes from the letters T and ϕ, the latter as a symbol of entropy. As we saw in Chapter III, Section 9, the ordinate variable $\ln \theta$ is proportional to the specific entropy, so that this diagram can also be considered as having coordinates s, T.

Although the ordinates are proportional to $\ln \theta$, the adiabats (straight lines of constant θ) are labelled with the values of θ; i.e., the vertical axis is a logarithmic scale of θ. The diagram is represented schematically in Figure VI-2.

---

* We shall mention first the ordinate and then the abscissa. This convention is opposite to the general use in mathematics, but is frequently applied in thermodynamics.

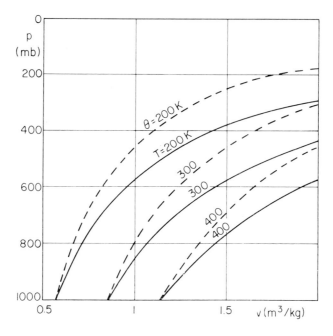

Fig. VI-1. Clapeyron diagram.

As for properties of the tephigram:

(a) The angle between adiabats and isotherms is equal to 90°, a very favorable condition.

(b) Isotherms and adiabats are straight lines. From the equation of adiabats

$$\theta = T\left(\frac{1000}{p}\right)^{\varkappa_d} \tag{1}$$

(where p is expressed in mb), we have

$$(\ln \theta) = \ln(T) - \varkappa_d \ln p + \text{const.} \tag{2}$$

where the brackets indicate which are the variables of the diagram. Isobars ($p = $ constant) are therefore logarithmic, but their curvature is small within the usual range.

(c) It is equivalent. This can be shown by considering the work a performed on a unit mass of air in a reversible cycle

$$a = -q = -\oint T \, ds. \tag{3}$$

Introducing (Chapter III, Equation (54)): $ds = c_p \, d \ln \theta$, we obtain:

$$a = -c_p \oint T \, d \ln \theta = -c_p \Sigma_{\text{teph}} \tag{4}$$

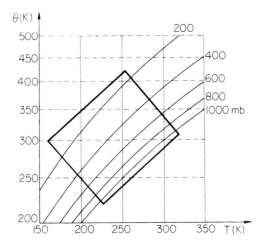

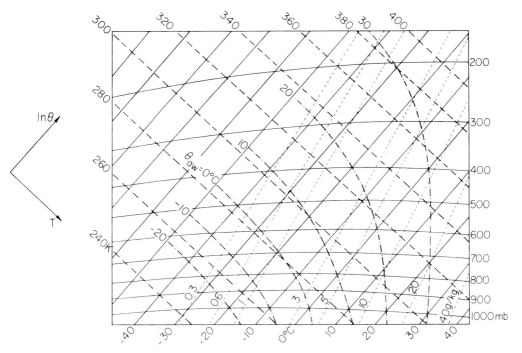

Fig. VI-2. Tephigram.

where $\sum_{\text{teph}}$ is the area determined by the cycle on the diagram, taken as positive if described counterclockwise, as the integration has been performed taking the vertical coordinate as the independent variable. If the air is dry, $c_p = c_{p_d}$.

(d) From Equation (2) it is obvious that isobars are congruent with respect to a displacement along the ordinate. Thus the isobar kp may be obtained by displacing vertically the isobar p by the constant quantity $\varkappa_d \ln k$. This property could be used to represent two different intervals of p and θ with the same set of curves (see Section 5).

(e) To comply with this criterion, the zone of interest in the atmosphere (shown in a rectangle at the top of Figure VI-2) is frequently represented, after a clockwise turn of about 45°, as in the lower part of Figure VI-2. Thus isobars approximate horizontal straight lines. We shall frequently use this arrangement in our schematic diagrams of the following chapters.

6.4. Curves for Saturated Adiabatic Expansion. Relative Orientation of Fundamental Lines

We consider now the isotherms and the equisaturated lines on the tephigram. Let us assume that, starting from an image point P, we go up along the corresponding isotherm T to point P', which has the pressure $p' < p$ (see Figure VI-3). Our system will consist of a certain mass of saturated air, and will be closed. Its mixing ratio r_w will therefore remain constant, unless condensation occurs. The first formula in Chapter IV, Equation (79)

$$r_w = \frac{\varepsilon e_w(T)}{p - e_w(T)}$$

shows that the saturation mixing ratio at the new pressure, r'_w, exceeds r_w, as $e_w(T)$, which is only a function of T, remains constant, while the pressure p is lower. Therefore, there was no condensation and, on the contrary, the air has become unsaturated. The saturation mixing ratio also increases if we increase T at constant p. We conclude that the equisaturated line must be inclined to the left of the isotherm, as shown in Figure VI-3, in such a way that the decrease in $e_w(T)$ must compensate the decrease in p (in order to keep a constant mixing ratio).

Let us consider now a new process, the adiabatic ascent of saturated air, for the sole purpose of describing the diagrams. Analytical consideration of this process will be given in Chapter VII, Sections 8 and 9; here we shall only consider it qualitatively. The ascent of the air implies an expansion, and being an adiabatic process, the temperature decreases and water vapor condenses. The curve describing this process

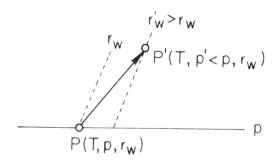

Fig. VI-3. Relative orientation of isotherms and equisaturated lines.

does not follow the non-saturated adiabat any more, because the latent heat of vaporization released during the condensation is absorbed by the air, with the effect that there is a smaller drop in temperature for the same decrease in pressure. Therefore this curve, when considered from point P up, is situated to the right of the dry adiabat.

On the other hand, the saturated adiabat must cut, in the direction of expansion, equisaturated lines of lower and lower values of r_w, in view of the fact that the mixing ratio of the air diminishes continuously with condensation, and in every point must be equal to that of the equisaturated line passing through it (because the air is saturated). This requires that the inclination of the saturated adiabat should be such that the curve lies, for every point, between the dry adiabat and the equisaturated line.

Finally, it is obvious that an isobaric cooling must be beyond all these curves, which all imply a decrease in pressure. We conclude, therefore, that the relative orientation of the fundamental lines must be that indicated in Figure VI-4.

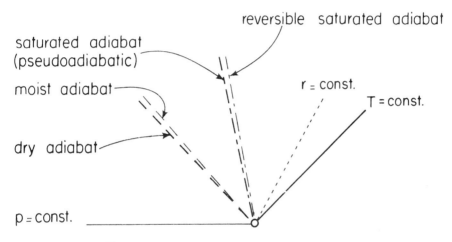

Fig. VI-4. Relative orientation of fundamental lines.

An adiabat corresponding to moist air will show a variation of temperature with pressure slightly smaller than that for dry air (because $\varkappa < \varkappa_d$; see Chapter IV, Section 12, Equation (89)), and its slope will depend on the specific humidity; its orientation will be such as indicated with a thin dashed curve in Figure VI-4. Only dry adiabats are printed on diagrams.

We must also remark that the saturated adiabats printed on the diagrams correspond to the process in which all the water goes out of the system as soon as it is condensed (pseudo-adiabatic process; Chapter VII, Section 9). If we draw the curves corresponding to the process in which all the water remains in the system (reversible saturated adiabatic process; Chapter VII, Section 8), they should show a slightly smaller slope, because the liquid water will also lose heat while cooling. This is shown in Figure VI-4 by the thin dash-and-dot curve.

This sequence in the orientation of the fundamental lines holds for all diagrams.

As in the previous figures, we shall always use the following convention to represent the fundamental lines: full lines for isotherms and isobars, dashed curves for dry adiabats, dash-and-dot curves for saturated adiabats and dotted (or short-dashed) curves for equisaturated lines. Values of pressures are given in mb, and values of mixing ratios in g kg^{-1}.

6.5. Emagram or Neuhoff Diagram

We consider now the diagram with coordinates $-\ln p$ (logarithmic scale of p), T (Figure VI-5). It is usually called the *emagram*, following Refsdal, from 'energy per unit *ma*ss dia*gram*', although this name could also be applied to all equivalent diagrams in general. Regarding its main properties:

(a) The angles depend on the scales used for the coordinate axes and vary with the point; the scales are usually chosen so that they are about 45°.

(b) Isobars and isotherms are straight lines. Adiabats can be written (cf. Equation (1)) as

$$(-\ln p) = -\frac{1}{\varkappa_d}\ln(T) + \frac{1}{\varkappa_d}\ln\theta + \text{const.} = -\frac{1}{\varkappa_d}\ln(T) + \text{const.} \qquad (5)$$

where the two coordinate variables are between brackets; this shows that adiabats are logarithmic curves (normally with a small curvature).

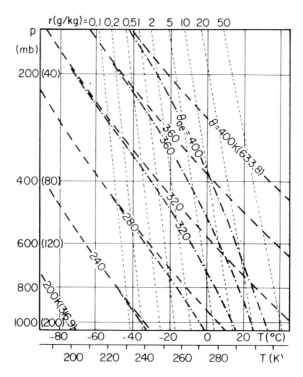

Fig. VI-5. Emagram.

(c) It may be shown that it is an equivalent diagram. We have:

$$\delta a = - p\, dv = - d(pv) + v\, dp = - R\, dT + v\, dp.$$

In a cycle

$$a = - R \oint dT + \oint v\, dp = \oint v\, dp = - R \oint T\, d(- \ln p) = - R \sum_{em} \quad (6)$$

where $\sum_{em}$ stands for the area enclosed in the cycle, positive if described counter-clockwise. For dry air, $R = R_d$.

(d) Adiabats are congruent with respect to a displacement along the vertical axis. This can be seen from Equation (5), taking into account that $\ln(kp) = \ln p + \ln k = \ln p +$ + const. This property can be used to represent two different intervals of pressures and potential temperatures with the same set of curves. Thus the same curve represents p as a function of T for θ = const. and (kp) as a function of T for $(k^{-\kappa_d}\theta)$, as can be verified by adding $- \ln k$ on both sides of Equation (5); this is done in Figure VI-5 for a value $k = 1/5$, with some values of (kp) and $(k^{-\kappa_d}\theta)$ indicated between brackets.

(e) Ordinates are roughly proportional to heights, as will be seen in Chapter VIII, Section 11.

In order to increase the angle between isotherms and adiabats, thereby making the diagram more sensitive to changes of slope of the curves $T = f(p)$, a *skew emagram* is also used with the axes at an angle of 45°, as indicated in Figure VI-6. Comparison with Figure VI-2 shows that the skew emagram is very similar to the tephigram, with straight isobars and slightly-curved adiabats replacing slightly-curved isobars and straight adiabats.

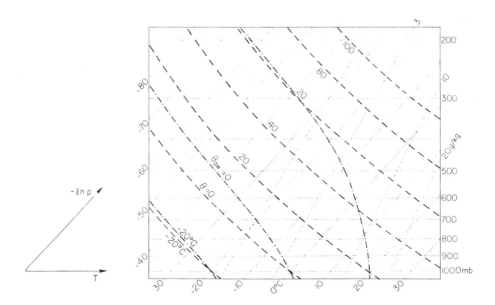

Fig. VI-6. Skew-emagram.

6.6. Refsdal Diagram

The diagram of Refsdal has coordinates $-T\ln p$, $\ln T$, labelled with the values of p and T, respectively. Regarding some of its properties:

(a) Only isotherms are straight lines.

(b) The angle between adiabats and isotherms depends on the scales.

(c) It is equivalent. Integrating with respect to the abscissa:

$$\int (-T\ln p)\,d\ln T = -\int \ln p\, dT = -T\ln p + \int T\,d\ln p. \qquad (7)$$

In a cycle the first term cancels and the area becomes (changing sign for a cycle described counterclockwise):

$$\Sigma_{\text{Ref}} = -\oint T\,d\ln p = \oint T\,d(-\ln p). \qquad (8)$$

Therefore, for a cycle (cf. Equation (6)):

$$a = -R_d \Sigma_{\text{Ref}}. \qquad (9)$$

(d) Neither adiabats nor isobars are congruent for displacement along the axis.

(e) As will become clear in Chapter VIII, Section 11, the vertical coordinate is essentially proportional to height in the atmosphere.

Figure VI-7 represents a simplified Refsdal diagram.

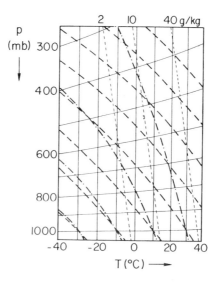

Fig. VI-7. Refsdal diagram.

6.7. Pseudoadiabatic or Stüve Diagram

The pseudoadiabatic or Stüve diagram, shown schematically in Figure VI-8, has the coordinates $-p^{\varkappa_d}$, T, the ordinates being labelled with the values of p. The region of the appropriate range of the variables which is usually represented corresponds to the rectangle in thick lines shown in the figure.

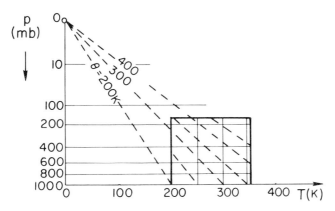

Fig. VI-8. Pseudoadiabatic or Stüve diagram.

Some of its properties are:

(a) The angle between adiabats and isotherms depends on the scales and is normally around 45°.

(b) Isotherms and isobars are straight lines, as well as adiabats, as becomes obvious by considering Poisson's equations (Chapter II, Section 7). The adiabats converge to the point $p=0$, $T=0$, as shown in the figure.

(c) It is not equivalent. However, energies corresponding to equal areas do not differ greatly; for instance, 1 cm² at 400 mb represents about 25% more energy than at 1000 mb.

6.8. Area Equivalence

We have seen already the area-energy equivalence properties of the diagrams described, as particular cases. Here we shall describe a general method which may be applied to check that property on any diagram.

Let us call Σ the area enclosed by a certain contour C in the x, y plane. We shall assume that these two variables are related to another pair u, w, that there is a one-to-one correspondence between the points in the plane x, y and those in u, w, and that to the points in Σ correspond those of a surface Σ' enclosed by a contour C'.

The area of an elementary surface $d\Sigma$ in x, y may be expressed by the cross product of two vectors $d\mathbf{x}$ and $d\mathbf{y}$:

$$d\Sigma = d\mathbf{x} \times d\mathbf{y}.$$

Similarly

$$d\Sigma' = d\mathbf{u} \times d\mathbf{w}$$

(see Figure V-9).

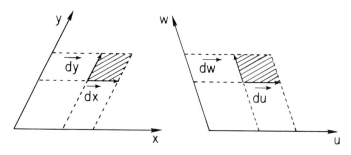

Fig. VI-9. Area equivalence.

Every point can be represented in each plane by two vectors; in the x, y plane: $x = x(u, w)$, $y = y(u, w)$, where the functions represent the relation between the points in both planes. Differentiation gives:

$$d\mathbf{x} = \frac{\partial x}{\partial u} d\mathbf{u} + \frac{\partial x}{\partial w} d\mathbf{w}$$

$$d\mathbf{y} = \frac{\partial y}{\partial u} d\mathbf{u} + \frac{\partial y}{\partial w} d\mathbf{w}$$

and

$$d\Sigma = \left(\frac{\partial x}{\partial u} d\mathbf{u} + \frac{\partial x}{\partial w} d\mathbf{w}\right) \times \left(\frac{\partial y}{\partial u} d\mathbf{u} + \frac{\partial y}{\partial w} d\mathbf{w}\right)$$

$$= \left(\frac{\partial x}{\partial u}\frac{\partial y}{\partial w} - \frac{\partial x}{\partial w}\frac{\partial y}{\partial u}\right)(d\mathbf{u} \times d\mathbf{w})$$

where we have taken into account that $d\mathbf{u} \times d\mathbf{u} = d\mathbf{w} \times d\mathbf{w} = 0$ and $d\mathbf{u} \times d\mathbf{w} = -d\mathbf{w} \times d\mathbf{u}$. Therefore

$$d\Sigma = J d\Sigma' \qquad (10)$$

where

$$J = \begin{vmatrix} \dfrac{\partial x}{\partial u} & \dfrac{\partial x}{\partial w} \\ \dfrac{\partial y}{\partial u} & \dfrac{\partial y}{\partial w} \end{vmatrix} = \frac{\partial x}{\partial u}\frac{\partial y}{\partial w} - \frac{\partial x}{\partial w}\frac{\partial y}{\partial u} = J\frac{(x, y)}{(u, w)}.$$

J is called the Jacobian of the co-ordinate transformation.

The condition for the transformation to be area-preserving is that the Jacobian be equal to unity or to a constant. In the last case the area in the new co-ordinates will be proportional to that in the first system. If in the case of aerological diagrams, $\sum$ is proportional to the energy, $\sum'$ will also be proportional to it, but with a different proportionality constant, as shown by Equation (10), integrated to

$$\sum = J \sum'. \tag{11}$$

In order to see if a diagram is area-equivalent, it will be sufficient to calculate the Jacobian that relates it to Clapeyron's diagram (or to any other area-equivalent diagram). The work performed on the system per unit mass in a process will be proportional to the area determined by the process if J is constant, and J will be the proportionality constant.

As an example, we shall demonstrate the area equivalence of the emagram, starting from Clapeyron's diagram.

We have to transform the coordinate system form $x = v$, $y = -p$ to $u = T$, $w = -\ln p$.

$$x = v = \frac{RT}{p} = Rue^w$$

$$y = -p = -e^{-w}$$

$$\frac{\partial x}{\partial u} = Re^w \qquad \frac{\partial x}{\partial w} = Rue^w$$

$$\frac{\partial y}{\partial u} = 0 \qquad \frac{\partial y}{\partial w} = e^{-w}$$

$$J = R. \tag{12}$$

If we take into account that the integrations are performed in such a way that the area is described clockwise in the Clapeyron diagram and counterclockwise in the emagrams, we must add a negative sign. As the area in the Clapeyron diagram gives the work a, this result, introduced into Equation (11), reproduces formula (6).

6.9. Summary of Diagrams

The main properties of the diagrams mentioned are summarized in the following table.

Diagram	Abscissa	Ordinate	J (from $-p, v$)	Straight lines			Angle adiabats-isotherms
				Isobars	Adiabats	Isotherms	
Clapeyron	v	$-p$	1	yes	no	no	small
Tephigram	T	$\ln \theta$	c_p	no	yes	yes	90°
Emagram	T	$-\ln p$	R	yes	no	yes	~45° (unless skewed)
Refsdal	$\ln T$	$-T \ln p$	R	no	no	yes	~45°
Pseudo-adiabatic or Stüve	T	$-p^{x_d}$	$\neq$ const	yes	yes	yes	~45°

6.10. Determination of Mixing Ratio from the Relative Humidity

In Chapter IV, Equations (85) and (86) we have seen the expressions to compute r from U_w, which is generally the parameter directly obtained from observations. A simple rule (Refsdal) allows one to determine it conveniently on the diagrams. The procedure is:

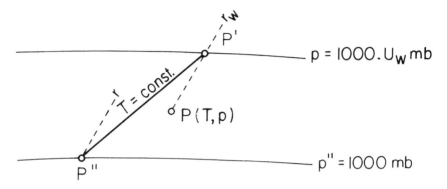

Fig. VI-10. Determination of mixing ratio from the relative humidity.

(1) From the image point $P(T, p)$ (see Figure VI-10) follow the vapor line passing through it up to its intersection P' with the isobar $p = 1000\, U_w$ mb.

(2) Follow the isotherm passing through P' down to its intersection with the 1000 mb isobar (P''). The value of the vapor line passing through P'' gives the mixing ratio r of the air.

This may be demonstrated using the approximate formulae of Chapter IV, Section 11:

$$r = \frac{\varepsilon e_w(T')}{p''} = \frac{\varepsilon U_w e_w(T')}{p'} = U_w r_w(T', p') = U_w r_w(T, p) \tag{13}$$

where the first equation comes from applying the general formula $r_w \cong \varepsilon e_w/p$ to P'' (r being the saturation value at that point), the second expression is obtained by substituting $p'' = p'/U_w$, and the third expression from applying $r \cong \varepsilon e_w/p$ to P'.

6.11. Area Computation and Energy Integrals

We have seen that in area-equivalent diagrams, the area enclosed by a cyclic process is proportional to the work a and to the heat $q = -a$ received by unit mass of air during the process. We are also interested in the values of these quantities for any open processes. We shall consider such processes for moist air, when no condensation takes place. In that case, the variation in state functions will no longer be zero in general. We shall have:

$$\Delta u = c_v \Delta T \tag{14}$$

$$\Delta h = c_p \Delta T \tag{15}$$

$$\Delta s = c_p \ln \frac{\theta_2}{\theta_1} \tag{16}$$

where the subscripts 1, 2 refer to the initial and final states. ΔT can be read directly. Δs can be computed from values of θ_1 and θ_2 or by measuring $\Delta \ln \theta = \ln(\theta_2/\theta_1)$ on a tephigram with an appropriate scale in entropy units.

The calculation of the heat q and the work a, which are not state functions, requires a knowledge of the process undergone by the system. If this process is plotted on a diagram, the calculation can be simplified by graphical approximations. Let us consider the following expressions:

$$a = -\int_1^2 p \, dv = -pv\big|_1^2 + \int_1^2 v \, dp = -R\Delta T + R \int_1^2 T \, d \ln p \tag{17}$$

$$q = \int_1^2 T \, ds = c_p \int_1^2 T \, d \ln \theta. \tag{18}$$

If we can calculate either q or a, the other one follows immediately from the first principle

$$\Delta u = a + q \tag{19}$$

and Equation (14).

The integral term in the last expression of a in Equation (17) is directly related to an emagram area. If we call $\sum_{em}$ the area 1-2-3-4-1 in Figure VI-11 (described counterclockwise), where the curve 1-2 represents the process, we have

$$\sum_{em} = \int_1^2 T \, d(-\ln p) = -\int_1^2 T \, d \ln p \tag{20}$$

so that

$$a = -R(\Delta T + \sum_{em}) \tag{21}$$

and (from Equations (14) and (19))

$$q = c_p \Delta T + R \sum_{em}. \tag{22}$$

It is obvious that $\sum_{em}$ can be computed by drawing the mean temperature $\bar{T}$ that compensates areas (1'-2' in the figure) and taking the area of the rectangle 1'-2'-3-4-1'. This is the *method of the mean isotherm*. It will be

$$\sum_{em} = -\bar{T} \ln \frac{p_2}{p_1}. \tag{23}$$

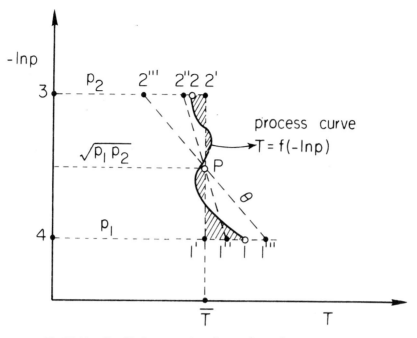

Fig. VI-11. Graphical computation of energy integrals on an emagram.

If we now substitute any straight line, such as $1''$-$2''$, passing through the middle point P in the isotherm, for $1'$-$2'$, this will determine the same area. Therefore, reversing the argument, we could choose any straight line following as closely as possible the curve 1-2 and compensating areas (which will be much easier to do with a good approximation); the middle point P of this straight line between the isobars will give $\bar{T}$. This is illustrated in Figure VI-11 by the line $1''$-$2''$; it will be $\bar{T} = (T_1'' + T_2'')/2$, where T_1'' and T_2'' are the temperatures corresponding to the points $1''$ and $2''$. We notice that the isobar passing through P has the value $(p_1 p_2)^{1/2}$.

The area $\sum_{em}$ can also be computed by the *method of the mean adiabat*. This consists in finding the dry adiabat $1'''$-$2'''$ that passes through 1-2, compensating areas. If p_1 and p_2 are not very distant, $1'''$-$2'''$ can be taken on an emagram as a straight line, in which case it should pass through P. Now $\sum_{em}$ can be computed from the area $1'''$-$2'''$-3-4-$1'''$; this gives

$$\sum_{em} = -\int_1^2 T \, d\ln p = -\frac{1}{R}\int_1^2 v \, dp = -\frac{1}{R}\int_{1'''}^{2'''} v \, dp \qquad (24)$$

where the last integral is taken along $1'''$-$2'''$, i.e., corresponds to an adiabatic process for which $\delta q = 0$. Therefore, from the first principle:

$$\int_1^2 v \, dp = \Delta h - q = \Delta h = c_p(T_2''' - T_1''') \qquad (25)$$

and

$$\sum_{em} = \frac{1}{\varkappa}(T_1''' - T_2''') \qquad (26)$$

where T_1''' and T_2''' are the temperatures at which the area-compensating adiabat intersects the isobars p_1 and p_2.

Wherever the product RT appears in the formulas, it can be substituted by the equivalent $R_d T_v$ (Chapter IV, Section 11), coupling the use of the dry air constant R_d with that of the virtual temperature; thus Equation (21) can be written with R_d, ΔT_v and $\sum_{em}$ given by Equation (23) with $\overline{T}_v$ rather than $\overline{T}$. On the other hand, $c_p T \neq c_{p_d} T_v$, so that a similar substitution could not be made in Equation (25).

Equations (23) and (26) provide convenient methods for computing energy integrals on an emagram. Now we shall consider the use of the tephigram. The heat q can be expressed by

$$q = \int_1^2 T\,ds = c_p \int_1^2 T\,d\ln\theta. \qquad (27)$$

The last integral is clearly related to the tephigram. If we consider Figure VI-12, and call $\sum_{te}$ the area 1-2-3-4-1 (1-2, as before, represents the process),

$$\sum_{te} = \int_1^2 T\,d\ln\theta. \qquad (28)$$

As for the emagram, we can apply the method of the mean isotherm (1'-2' in the figure) and write

$$\sum_{te} = \overline{T}\ln\frac{\theta_2}{\theta_1} = \varkappa\overline{T}\ln\frac{p_1'}{p_2'} = \frac{1}{c_p}\overline{T}\Delta s \qquad (29)$$

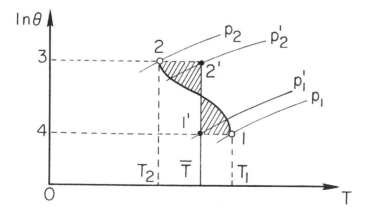

Fig. VI-12. Graphical computation of energy integrals on a tephigram.

where we have used Poisson's equation $\ln\theta = \ln T - \varkappa \ln p + \varkappa \ln 1000$ to relate the potential temperatures and the pressures along the isotherm $\overline{T}$, and the last expression indicates that a scale of entropies (rather than the logarithmic scale of θ) would provide a simple way of measuring the area.

It may be more convenient to work with the tephigram in a similar way as with the emagram – that is, integrating between isobars instead of between adiabats. We shall consider this procedure, which is not rigorous, but gives a good approximation provided the process occurs within a layer Δp which is not too thick.

We first consider that the pressure scale is logarithmic along an isotherm; for, by definition of θ,

$$\ln \theta = C - \varkappa \ln p \qquad (30)$$

where C is a constant at constant T, and as $\ln\theta$ is the ordinate variable, the distances between the isobars along T vary linearly with $\ln p$ (see Figure VI-13). It may be seen that, if we take the isobars as approximately straight, which will be a good approximation for small regions, the tephigram may be considered as a skew emagram, with

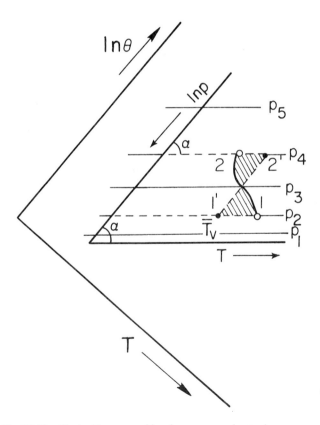

Fig. VI-13. The tephigram considered as an approximate skew-emagram.

an angle α between the axes of about $45°$. The area determined on this diagram is

$$\Sigma'_{te} = \left[\int_1^2 T\, d\ln p\right] \sin \alpha \tag{31}$$

which coincides with Σ_{te} except for the constant proportionality factor $\sin \alpha$. Within this approximation, we can therefore work with the tephigram as if it were an emagram, and apply the same methods as described before.

It should be noticed that the areas Σ_{em}, Σ_{te} have been defined with the points 3, 4 on the vertical axis (origin of abscissae, $T=0$), which is usually outside the region represented in the diagram.

We can also remark that the area compensation in the tephigram will be exactly as good a procedure as in the emagram whenever the lines enclosing the areas have the same meaning, because the area equivalence is preserved. Thus, for instance, the slightly curved adiabat in the emagram and the straight adiabat in the tephigram will compensate areas equally well, provided in each case the adiabat passes through the point determined by $p = (p_1 p_2)^{1/2}$ and the same $\bar{T}_v$.

These methods can also be used for the computation of heights, and this is their main application, which will be considered in Chapter VIII, Section 11.

PROBLEMS

1. Construct a tephigram, drawing the following isopleths:
    $\theta = 250, 270, 290, 310, 330, 350\,\text{K}$ – Scale: 1 unit of $\log_{10} = 100$ cm.
    $T = 230, 250, 270, 290, 310\,\text{K}$ – Scale: $10°\text{C} = 1.5$ cm.
    $p = 1000, 900, 800, 700, 600, 500$ mb.
    $r_w = 1, 5, 10$ g kg^{-1}.
    (For saturated adiabats, cf., Problem VII-8).
2. Starting from the emagram, and using the Jacobian, show that the tephigram is also an area-equivalent diagram, and that the proportionality factor between the area and the energy is c_p.
3. Show with the Jacobian that Stüve's diagram is not area-equivalent.
4. An air mass is defined by $T=20.0°\text{C}$, $p=900$ mb, $U_w=70\%$. Find the following parameters on a tephigram: r, r_w, θ, T_d.
5. Assume that dry air undergoes a process which can be described by a straight line on a tephigram, going from $(10°\text{C}, 1000\text{ mb})$ to $(0°\text{C}, 850\text{ mb})$. Compute Δu, Δh and Δs, and apply the graphical methods of the mean temperature and of the mean adiabat to compute a and q.

CHAPTER VII

THERMODYNAMIC PROCESSES IN THE ATMOSPHERE

In this chapter we shall study a number of thermodynamic processes of great importance in the atmosphere.

Two of these processes will be isobaric cooling (Sections 1 and 2). We have in that case

$$\delta q \neq 0; \quad dp = 0; \quad dh = \delta q.$$

All the other processes to be analyzed will be adiabatic, except that of Section 11. Three of them will also be isobaric; for them:

$$\delta q = 0; \quad dp = 0; \quad dh = \delta q + v\, dp = 0.$$

That is, they will be isenthalpic. And they will not in general be reversible. These processes are the vaporization or condensation of water in non-saturated air (Section 3), the horizontal mixing of two air masses (Sections 4 and 5) and the freezing of a cloud (Section 10). The vertical mixing of air masses (Section 12) may be considered as a combination of two adiabatic processes, one of them isobaric. Of the remaining processes, one will be non-adiabatic: the polytropic expansion (Section 11); another will be adiabatic and reversible, and therefore isentropic: the reversible expansion of saturated air (Sections 6 – 8). The last one will be performed with an open system: the pseudoadiabatic expansion (Section 9).

It is easy to understand why most of the important processes in the atmosphere are adiabatic. Our systems will in general be rather large portions of the atmosphere. We can generally disregard what happens on the ground surface and the cooling by radiation, and the heat conduction processes through the air are relatively inefficient. If the air parcel is large enough to be insensitive to what happens on its borders, we may consider it as a closed system which does not exchange heat with its surroundings.

7.1. Isobaric Cooling. Dew and Frost Points

In every closed system consisting of moist air, the specific humidity and the mixing ratio remain constant. This is not so for the partial pressure of water vapor or for the relative humidity. The first one is constant with temperature but proportional to the total pressure (cf. Chapter IV, Equations (76); also by considering $e = N_v p$). The relative humidity varies strongly with the temperature, due to the rapid variation of the saturation vapor pressure e_w, and is proportional to the pressure, through e.

Let us consider a mass of moist air cooling at constant pressure, and therefore contracting. The variables q, r and e will remain constant, but U_w will increase due to the decrease of e_w. If the cooling continues, e_w will become equal to e, and U_w will reach unity: the air has reached saturation. The temperature at which saturation is reached is called the *dew point*, T_d. This is a new variable that can be used to characterize the humidity of the air. If the saturation is reached with respect to ice, rather than to water (at a temperature below 0 °C), the temperature is called the *frost point*, T_f.

Now, let us assume that the pressure may change (by rising or subsiding of the air) and that the humidity of our system may also change (for instance, by incorporating water vapor by turbulent diffusion from a water surface, or because rain fell through the air mass), or that we simply want to compare air masses that differ in humidity and pressure. If we want to find the relation between the dew point and the mixing ratio and total pressure, we must apply the Clausius-Clapeyron equation describing the equilibirum curve water-vapor. We first differentiate logarithmically the approximate expression $e \cong pr/\varepsilon$ and obtain

$$\mathrm{d}\ln e \cong \mathrm{d}\ln r + \mathrm{d}\ln p. \tag{1}$$

The Clausius-Clapeyron equation is

$$\frac{\mathrm{d}\ln e}{\mathrm{d}T_d} = \frac{l_v}{R_v T_d^2} \tag{2}$$

where e is the vapor pressure in our air mass at temperature T. T_d is the dew temperature, and therefore corresponds to e over the saturation curve (see Figure VII-1). Due to this relation, T_d and e are humidity parameters giving equivalent information. Solving for $\mathrm{d}T_d$, we find

$$\mathrm{d}T_d = \frac{R_v T_d^2}{l_v} \mathrm{d}\ln e \cong \frac{R_v T_d^2}{l_v} (\mathrm{d}\ln r + \mathrm{d}\ln p). \tag{3}$$

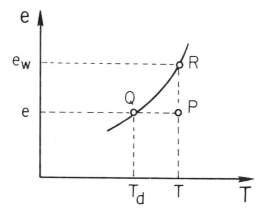

Fig. VII-1. Relation between temperatures and vapor pressures.

Or, if we want to express it as a relative variation $dT_d/T_d = d \ln T_d$:

$$d \ln T_d = \frac{R_v T_d}{l_v} d \ln e \cong \frac{R_v T_d}{l_v} (d \ln r + d \ln p) \cong 5 \times 10^{-2} (d \ln r + d \ln p) \quad (4)$$

where the last expression corresponds to the approximation $T_d \sim 270 \, \text{K}$ and indicates that the relative increase in T_d is about 5% of the sum of relative increases in r and p.

By integrating Equation (2) between T_d and T, we obtain

$$\ln \frac{e_w}{e} = -\ln U_w = \frac{1}{R_v} \int_{T_d}^{T} \frac{l_v}{T^2} dT \cong \frac{l_v}{R_v} \frac{T - T_d}{T T_d} \quad (5)$$

and solving for $(T - T_d)$, using decimal logarithms and substituting numerical values for the constants, we have

$$T - T_d = 4.25 \times 10^{-4} \, T T_d (-\log U_w), \quad (6)$$

which, for a rough estimate of $(T - T_d)$ as a function of U_w, can be written, with $T T_d \cong 290^2$:

$$T - T_d \cong 35(-\log U_w). \quad (7)$$

Figure VII-1 shows the relations between the values for temperature and vapor pressure. The above integration was performed between Q and R.

So far we have considered the temperatures at which the equilibrium curve of phase transition is reached, such as T_d for point Q in Figure VII-1 or point D in Figure VII-2, and T_f for point F in Figure VII-2. But no thermodynamic argument can

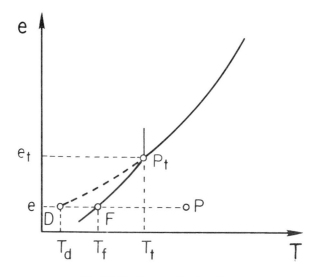

Fig. VII-2. Dew-and frost-points.

say whether the processes of condensation or sublimation will actually start taking place at these points, when isobaric cooling proceeds. Condensation does start at the dew point, although this process requires the presence of solid surfaces or of certain particles, called condensation nuclei. If neither of them were present, there would not be condensation (at least not along a certain temperature interval below the dew point) and the air would become *supersaturated* in water vapor (with $e > e_w$). However, atmospheric air always contains abundant condensation nuclei, and supersaturation does not occur to any appreciable extent. Therefore, although this is an important subject in the study of cloud physics, we do not need to be concerned with it here.

The situation is somewhat different regarding the processes of sublimation and of freezing. If ice surfaces are already present, sublimation or freezing will proceed readily on these surfaces as soon as the water or the vapor, as the case may be, reaches the equilibrium curve. In the absence of ice surfaces, on the other hand, although certain surfaces, either in macroscopic extensions or on minute particles of the atmospheric aerosol ("ice nuclei") favor the appearance of ice crystals, they only become active at temperatures well below the equilibrium curve. Neither does spontaneous nucleation take place with small supercooling of water or supersaturation of water vapor.

Therefore, when isobaric cooling of moist air proceeds, starting from a point such as P in Figure VII-2, sublimation will not in general occur at F. Between F and D, air will be supersaturated with respect to ice, and may only condense to water at the point D. It is interesting to derive the relation between dew and frost temperatures. This is an important point with respect to aircraft icing and surface fogs.

If we apply the Clausius-Clapeyron Equation (Chapter IV, Equation (50)) to both the vaporization and the sublimation curves, between the points D and F and the triple point P_t, we obtain:

$$\ln \frac{e_t}{e} = \frac{l_v}{R_v} \frac{T_t - T_d}{T_t T_d} = \frac{l_s}{R_v} \frac{T_t - T_f}{T_t T_f} \tag{8}$$

where e_t is the triple-point vapor pressure, and l_v and l_s are considered as constants. Taking into account that $T_t = 273.16 \text{ K} \cong T_0 = 273.15 \text{ K}$, that the dew and frost points, expressed in °C are

$$t_d = T_d - T_0 \cong T_d - T_t \tag{9}$$

$$t_f = T_f - T_0 \cong T_f - T_t \tag{10}$$

and that

$$T_t T_d \cong T_t T_f \tag{11}$$

we may write

$$\frac{t_d}{t_f} \cong \frac{l_s}{l_v}. \tag{12}$$

Although the mean values to be taken for the latent heats vary somewhat with the temperature interval, we may use with good approximation the values for $-10\,°C$, which give

$$l_s/l_v \cong 9/8. \tag{13}$$

From this value, it is easily seen that

$$|t_d - t_f| \cong |t_d|/9 \cong |t_f|/8. \tag{14}$$

The dew point temperature can be found in a diagram by following the isobar from the image point P representing the air until it intersects the saturation mixing ratio line corresponding to the mixing ratio r of the air, as shown in Figure VII-3.

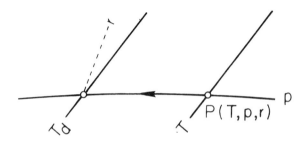

Fig. VII-3. Dew-point on a diagram.

We finally remark that, according to their definitions, T_d and T_f are (as e) invariants for isobaric changes of temperature of a closed system consisting of moist air.

7.2. Condensation in the Atmosphere by Isobaric Cooling

Dew and *frost* form as a result of condensation or sublimation of water vapor on solid surfaces on the ground, which cool during the night, by radiation, to temperatures below the dew or the frost point of the air in contact with them.

If a mass of atmospheric air cools isobarically until its temperature falls below the dew point, condensation will occur as microscopic droplets formed on condensation nuclei; we call this a *fog*. This occurs in the atmosphere due to the radiative cooling of the air itself or of the ground with which it is in contact (*radiation fogs*). As the droplets form, and because these droplets behave as black bodies in the wavelengths in which they irradiate, the radiation emitted by the layer increases, which favors further loss of heat. Condensation may also occur when an air mass moves horizontally over the ground toward colder regions, and becomes colder itself by heat conduction to the ground (*advection fogs*). In both cases, the cooling is practically isobaric, since pressure variations at the surface are usually very small (on a relative basis).

Once condensation starts, the temperature drops much more slowly, because the heat loss is partially compensated by the release of the latent heat of condensation. This sets a virtual limit very close to the dew point, an important fact in forecasting minimum temperatures.

For an isobaric process, the heat absorbed is given by the increase in enthalpy (Chapter IV, Equation (106)):

$$\delta q = \mathrm{d}h \cong c_p \, \mathrm{d}T + l_v \, \mathrm{d}r. \tag{15}$$

If we write $r \cong \varepsilon e/p$, as p is a constant, $\mathrm{d}r \cong (\varepsilon/p) \, \mathrm{d}e$. In this case e corresponds to saturation, and we may apply the Clausius-Clapeyron Equation (Chapter IV, Equation (48)) and write $e = e_w$; we obtain:

$$\mathrm{d}r \cong \frac{\varepsilon}{p} \, \mathrm{d}e = \frac{\varepsilon l_v e_w}{p R_v T^2} \, \mathrm{d}T \tag{16}$$

and

$$\delta q = \left(c_p + \frac{\varepsilon l_v^2 e_w}{p R_v T^2} \right) \mathrm{d}T, \tag{17}$$

or else

$$\delta q = \left(\frac{c_p R_v T^2}{l_v e_w} + \frac{\varepsilon l_v}{p} \right) \mathrm{d}e_w. \tag{18}$$

The relation between $\mathrm{d}T$ and $\mathrm{d}e_w$ is indicated in Figure VII-4, on a vapor pressure diagram.

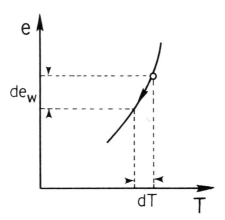

Fig. VII-4. Relation between the changes in temperature and in vapor pressure during condensation.

If we compute the heat loss $-\delta q$ from other data (e.g.: radiation loss), Equation (17) allows an estimation of the corresponding decrease in temperature $-\mathrm{d}T$. Similarly, the decrease in vapor pressure $-\mathrm{d}e_w$ may be computed from Equation (18). From the gas law, the mass of water vapor per unit volume is given by $e_w/R_v T$, and its variation with temperature is $\mathrm{d}e_w/R_v T - (e_w/R_v T^2) \, \mathrm{d}T$. Introducing the Clausius-Clapeyron relation between de_w and dT, it can be shown that the second term is much smaller than the first one ($\sim 5\%$) and can be neglected for the following approximate argument. We can write thus for the mass of condensed water per unit volume dc:

$$dc \cong -\frac{1}{R_v T} de_w = -\frac{l_v e_w}{R_v^2 T^3} dT \tag{19}$$

(where the last relation is given, as before, by the Clausius-Clapeyron equation). If, for instance, we want to know the necessary cooling to reach a concentration of liquid water $\Delta c = 1$ g water m^{-3} air, starting with saturated air at $10\,°C$, Equation (19) gives a result of $\Delta T = -1.6\,°C$ (where we substituted finite differences for differentials).

We may notice that according to Equation (19), and using Δ instead of d for the variations,

$$-\Delta e_w \cong R_v T \Delta c \tag{20}$$

is roughly constant for a constant Δc in the usual interval of temperatures. Computation gives 1.3 to 1.4 mb for $\Delta c = 1$ g water m^{-3} air. We might use the vapor pressure diagram T, e to represent lines of equal Δc. Their points would lie Δe_w above the saturation curve for each temperature, as shown schematically in Figure VII-5. These lines of constant liquid water content in fogs are related to visibility. Equation (18) shows that *for a given loss of heat $\delta q < 0$, $-de_w$ is larger if T is higher*, because e_w within the bracket increases more rapidly than T^2; dc, approximately proportional to $(-de_w)$ according to (19), is therefore larger for higher T and smaller for lower T. For this reason dense fogs are less frequent at low than at mild temperatures.

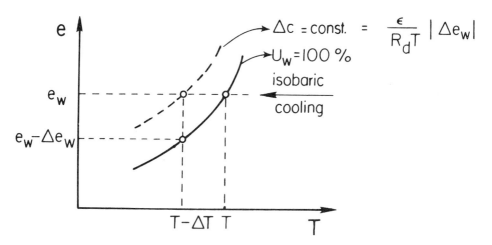

Fig. VII-5. Lines of constant liquid water content on a vapor pressure-temperature diagram.

7.3. Adiabatic Isobaric (Isenthalpic) Processes. Equivalent and Wet-Bulb Temperatures

We shall consider a closed system consisting of moist air and water. The same theory will hold for moist air and ice, with the corresponding substitutions (mass and specific heat of ice for mass and specific heat of water, latent heat of sublimation for latent heat of vaporization, etc.).

We shall study, for this system, adiabatic isobaric processes, therefore isenthalpic processes. We are interested in the expression for the enthalpy variation, which must be set equal to zero, thus providing an equation relating the temperature and humidity variables for these processes. The expression for enthalpy was calculated in Chapter IV, Section 14 under several forms; we shall use here Equation (104):

$$H = (m_d c_{p_d} + m_t c_w)T + l_v(T)m_v + \text{const.} \qquad (21)$$

If we consider two states of the system linked by an isenthalpic process, the above expression may be applied to one of them, and a similar expression to the other:

$$h' = (m_d c_{p_d} + m_t c_w)T' + l_v(T')m'_v + \text{const.}$$

where m_d, m_t and the additive constant are the same as for the former state. And from $h = h'$ we obtain:

$$(m_d c_{p_d} + m_t c_w)(T' - T) + l_v(T')m'_v - l(T)m_v = 0 \qquad (22)$$

which may be written:

$$T' + \frac{l_v(T')m'_v}{m_d c_{p_d} + m_t c_w} = T + \frac{l_v(T)m_v}{m_d c_{p_d} + m_t c_w}. \qquad (23)$$

The denominator on each side is a constant for each system. Each of the two sides of the equation is therefore a function of the state of the system only; in other words, this expression (either on the left or on the right side of Equation (23)) is an invariant for isenthalpic transformations. If we divide both numerator and denominator in the quotients by m_d, we have

$$T' + \frac{l_v(T')r'}{c_{p_d} + r_t c_w} = T + \frac{l_v(T)r}{c_{p_d} + r_t c_w} \qquad (24)$$

where $r_t = (m_v + m_w)/m_d = m_t/m_d$. Notice that $(c_{p_d} + r_t c_w)$, constant for each system, will vary for different systems according to their total contents in water substance (vapor plus liquid).

We may now start simplifying the expression by neglecting the heat capacity of the water:

$$T + \frac{l_v(T)r}{c_{p_d} + r c_w} \cong \text{const.}$$

Here the denominator is no longer a constant. If we also consider that $r c_w$ is small in comparison with c_{pd}, and neglect the temperature variation of l_v, Equation (24) becomes

$$T' + \frac{l_v}{c_p} r' \cong T + \frac{l_v}{c_p} r \cong \text{const.} \tag{25}$$

where c_p may be taken as c_{pd} or, with better approximation, as $c_{pd} + \bar{r} c_w$, where $\bar{r}$ is an average value of r (an even better approximation, from Equation (24), would be to use the maximum value of r, with all water as vapor).

Equations (23) and (24) do not imply any more approximations than those made in deriving formula (104) from Chapter IV, Section 14. Equation (25) could also have been derived directly from the approximate Equation (106) in that chapter.

Let us consider the physical process that links two specific states (T, r) – unsaturated moist air plus water – and (T', r') – saturated or unsaturated moist air without water – of the system, and to which Equation (24) corresponds. We have 1 kg of dry air with $10^3 r$ grammes of water vapor and $10^3 (r' - r)$ grammes of liquid water (which may or may not be as droplets in suspension). We are thus assuming that $r' > r$, and also that saturation is not reached at any time (except eventually when arriving at the final state). The liquid water evaporates, and so the mixing ratio increases from r to r'. As this water evaporates, it absorbs the vaporization heat, which must be provided by the moist air itself and the water, because we are assuming that the system is adiabatically isolated. The temperature decreases from T to T'. As at any instant the state of the system differs finitely from saturation, the process is a spontaneous and irreversible one. A process producing the opposite modification (condensation of water) could be imagined, but it would be actually impossible. This is however immaterial in our case, because at no stage have we made the assumption of reversibility and only the first law has been applied. We shall deal with such a process first.

The *isobaric equivalent temperature* or, more simply, the *equivalent temperature* is defined as the temperature that moist air would reach if it were completely dried by condensation of all its water vapor, the water being withdrawn in a continuous fashion; the whole process is performed at constant pressure and the system is thermally isolated (except for the removal of water). We shall designate it by T_{ie} or T_e. With this definition, the process previously described and Equation (24) will not be strictly applicable, due to the removal of liquid. If we want to make an exact formulation, we should start by considering the infinitesimal variation dH:

$$(m_d c_{pd} + m_v c_w) dT + l_v(T) dm_v = 0 \tag{26}$$

where we assume that there is no liquid water initially. dm_v is assumed to be negative, which means that a mass $|dm_v|$ of vapor condenses. Now, before considering further condensation, we remove that liquid water. The enthalpy of the system will thereby decrease by $h_w dm_v$, but this does not affect the value of T. For the next infinitesimal condensation, Equation (26) will again be valid, with the new value of m_v. Thus, the equation adequately describes the process; m_v and, therefore, the whole bracket in the first term are variable. An exact integrated expression could be obtained from (26) (see Problem 3). Within the usual approximation adopted in Equation (25), however, the term $m_v c_w$ is considered negligible, and we can write

$$T_{ie} = T + \frac{l_v}{c_{pd}} r \tag{27}$$

and introducing numerical values for l_v and c_{pd}:

$$T_{ie} = T + 2.5(10^3 \, r) \tag{28}$$

where $(10^3 r)$ is the mixing ratio expressed as (g of vapor kg^{-1} of dry air).

We may now consider the case when we go in a similar way from any value r to saturation, r_w. This is the process with which we are concerned in the use of the wet-bulb psychrometer. This instrument consists of two thermometers, one with a dry bulb to measure the air temperature, and the other one having its bulb covered with a wet muslin wick. Air must flow around the wet bulb and if it is non-saturated, water will evaporate until it becomes saturated. If steady state has been reached, the necessary enthalpy cannot come from the water, which has a constant temperature, but from the air itself. The thermodynamic system to be considered is a certain (any) mass of air that has flowed around the bulb, plus the mass of water which was incorporated into the air by evaporation from the muslin. We are thus considering a process similar to that which we have been studying, but with a difference: water has not undergone the same temperature variation as the initial air, as in the process of Equation (24), neither has it been incorporated and evaporated into the system at a variable temperature; in our present process the added water was from the beginning at the final temperature of the whole system. Again this difference is ignored in the approximation of Equation (25). We may then write, with the same approximation:

$$T_{ie} = T_{iw} + \frac{l_v}{c_{pd}} r_w = T + \frac{l_v}{c_{pd}} r, \tag{29}$$

where T_{iw} or T_w is the *isobaric wet-bulb temperature* or simply the *wet-bulb temperature*, and r_w is the saturation mixing ratio at the temperature T_{iw}. The wet-bulb temperature may thus be defined as the temperature which air attains when water is evaporated into it until saturation is reached, while the system (air plus water) is kept at constant pressure and does not exchange heat with the environment.

When Equation (29) is applied to compute T_{iw}, successive approximations must be made, as r_w is also unknown. In normal meteorological practice, however, the value of the isobaric wet-bulb temperature is known, and one wishes to deduce either the dew point, T_d, or the vapor pressure of the air, e, which is the saturation vapor pressure corresponding to the dew-point temperature. Replacing $r \cong \varepsilon e/p$ in (29) (right-hand equation), solving for e and considering that $e = e_w(T_d)$ (cf. Figure VII-1), we find

$$e_w(T_d) = e_w(T_{iw}) - \frac{c_{pd} p}{\varepsilon l_v}(T - T_{iw}). \tag{30}$$

This is one version of the so-called **Psychrometric Equation**. The coefficient of $p(T-T_{iw})$ is often referred to as the psychrometric constant, although it will vary slightly with T_{iw} (via the temperature dependence of l_v, and as a consequence of various second-order terms which have been ignored in Equation (30)).* At temperatures (T or T_{iw}) less than 0 °C, this equation still applies as long as the conventional wetted muslin evaporator does not freeze.

If we have ice (usually a very thin layer frozen on the thermometer bulb) as the evaporating phase, the equilibrium temperature is known as the *isobaric ice-bulb temperature*, T_{ii}. In this case, analogous to Equation (30),

$$e_w(T_d) = e_i(T_{ii}) - \frac{c_{pd}p}{\varepsilon l_s}(T - T_{ii}) = e_i(T_f), \tag{31}$$

where the actual vapor pressure can be interpreted in terms of either a dew-point or frost-point temperature (although the former is normally employed, as a conventional humidity parameter, in order to avoid ambiguity and apparent discontinuities in the moisture field).

The previous theory of the wet-bulb psychrometer does not take into account the possibility that part of the air flowing around the bulb may not reach saturation, or that the temperatures of air and water may not reach equilibrium. The identification of T_{iw} as defined by Equation (29) with the temperature read in the psychrometer is subject to experimental verification; this turns out to be satisfactory, provided there is enough ventilation**, and that the bulbs are small enough to exchange little infrared radiation with the screen or other objects at air temperature.

In regard to the process defining T_{iw}, the same observations as made for T_{ie} are pertinent. We deal with a process which is spontaneous (irreversible) in the sense of water evaporation; it would be an impossible one in the opposite sense.

Another important meteorological process is described by this type of transformation: air cooling by evaporation of rain, at a given level. In this case, the system consists of a certain air mass plus the water that evaporates into it from the rain that falls through it. The initial temperature of the water, in this case, will be that of the raindrops as they pass through that level, and the air may or may not reach saturation. If it does, its temperature will decrease to the value T_{iw}.

We may notice that T_{ie} and T_{iw} are linked by Equation (29), and they are the maximum and the minimum value, respectively, that the air may attain through the isenthalpic process that we have considered. They are therefore two parameters giving equivalent information about the temperature and humidity state of the air. This relation can be expressed graphically by representing the process on a vapor pressure

---

* Its value can be expressed by 0.000660 (1 + 0.00115 t_{iw}), where t_{iw} is the Celsius wet-bulb temperature.
** An air flow between 4 and 10 m s^{-1} over the bulbs is recommended.

diagram. If we substitute the approximate expression $\varepsilon e/p$ for r in Equation (25) and rearrange the latter, we obtain

$$e' - e = -\frac{c_{p_d} p}{\varepsilon l_v}(T' - T). \tag{32}$$

For each value of p, $(T' - T)$ is proportional to $(e' - e)$; i.e., the adiabatic isobaric process in which we are interested occurs in the diagram along a straight line passing through the image point $P(T, e)$ with a slope $-(c_{p_d}/\varepsilon l_v)p$. This is represented in Figure VII-6. Extending the line toward increasing temperatures, it will intersect the horizontal axis $(e=0)$ at T_{ie}, and in the other direction it intersects the saturation curve at T_{iw}.

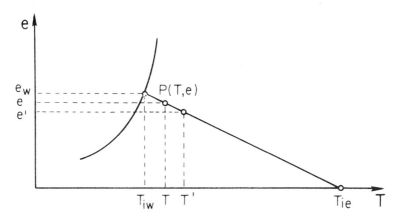

Fig. VII-6. Adiabatic isobaric process represented on a vapor pressure-temperature diagram.

7.4. Adiabatic Isobaric Mixing (Horizontal Mixing) Without Condensation

We shall now consider another adiabatic isobaric (and therefore isenthalpic) process: the mixing of two moist air masses, with different temperatures and humidities, but at the same pressure. This process corresponds in the atmosphere to horizontal mixing. Condensation is assumed not to take place.

If we use the subscripts 1 and 2 for the two masses, we shall have

$$m_1 h_1 + m_2 h_2 = (m_1 + m_2)h$$

or

$$\Delta H = m_1(h - h_1) + m_2(h - h_2) = m_1 \Delta h_1 + m_2 \Delta h_2 = 0, \tag{33}$$

where

$$\Delta h_1 = c_{p_1}(T - T_1)$$

$$\Delta h_2 = c_{p_2}(T - T_2);$$

T being the final temperature, and (Chapter IV, Equation (87))

$$c_{p_i} = c_{p_d}(1 + 0.87 q_i)$$

($i = 1, 2$). Substituting in ΔH:

$$m_1 c_{p_d}(1 + 0.87 q_1)(T - T_1) + m_2 c_{p_d}(1 + 0.87 q_2)(T - T_2) = 0. \tag{34}$$

And solving for T:

$$T = \frac{(m_1 T_1 + m_2 T_2) + 0.87(m_1 q_1 T_1 + m_2 q_2 T_2)}{m + 0.87(m_1 q_1 + m_2 q_2)}. \tag{35}$$

The total mass is $m = m_1 + m_2$, and the total mass of water vapor $m_1 q_1 + m_2 q_2$. As this must remain constant, it will be equal to mq, q being the final specific humidity. Therefore q is the weighted average of q_1 and q_2:

$$q = \frac{m_1 q_1 + m_2 q_2}{m}. \tag{36}$$

Introducing this relation in Equation (35) gives

$$T = \frac{(m_1 T_1 + m_2 T_2) + 0.87(m_1 q_1 T_1 + m_2 q_2 T_2)}{m(1 + 0.87 q)}.$$

If we neglect the water vapor terms, Equation (35) becomes:

$$T \cong \frac{m_1 T_1 + m_2 T_2}{m}. \tag{37}$$

That is, the final temperature is approximately given by the weighted average of the initial temperatures.

The potential temperature θ of the mixture is also given by similar formulas, as may be seen by multiplying both sides of Equation (35) or (37) by $(1000/p)^\kappa$, taking into account that p is a constant. In particular, we conclude that θ is also given approximately by the weighted average of θ_1 and θ_2:

$$\theta \cong \frac{m_1 \theta_1 + m_2 \theta_2}{m}. \tag{38}$$

If we use the approximate relation between e and q ($q \cong \varepsilon e/p$; see Chapter IV, Section 11) we obtain for the final vapor pressure:

$$e \cong \frac{m_1 e_1 + m_2 e_2}{m}. \tag{39}$$

We may notice that Equation (36) is valid for any mixture of two air masses, without condensation, independent of the pressure values or variations, while Equation (39) is only valid for an isobaric mixture, because in deriving it we assumed $p_1 = p_2 = p$.

Formulas (36), (37) and (39) indicate that the final q, T and e are obtained by computing the averages of their initial values, weighted with respect to the masses. Thus if P_1 and P_2 are the image points of the two air masses in a vapor pressure diagram (Figure VII-7), the image point P of the mixture will lie on the straight line joining P_1 and P_2, at a distance such that $P_1P/PP_2 = m_2/m_1$*.

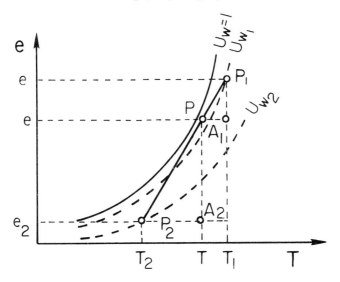

Fig. VII-7. Effect of mixing on relative humidity.

The dashed curves are curves of equal relative humidity ($U_{w1} = e_1/e_{w1}$ and $U_{w2} = e_2/e_{w2}$, respectively). It is easy to see that, due to the curvature of the relative humidity isopleths, the value of U_w of the mixture will always be higher than the weighted average of U_{w1} and U_{w2}; for instance, if $U_{w1} = U_{w2} = U_w$, any mixture will give a point to the left of the curve U_w.

7.5. Adiabatic Isobaric Mixing With Condensation

It becomes obvious on the vapor pressure diagram that, due to the increase in relative humidity produced by mixing, we may have the case of Figure VII-8, where the image point of the resulting mixture corresponds to a state of supersaturation, although both initial air masses were unsaturated.

As appreciable supersaturations cannot be realized in atmospheric air, condensation of water into droplets will occur in this case. Independently of how the real process occurs in nature, if we take into account that enthalpy changes do not depend on the

---

* This is easily seen by writing Equation (37) in the form

$$\frac{T - T_1}{T_2 - T} = \frac{m_2}{m_1}$$

and Equation (39) in a similar form, and considering the similarity of triangles PP_1A_1 and P_2PA_2 in Figure VII-7.

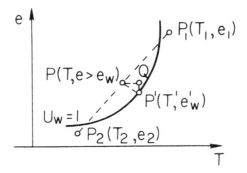

Fig. VII-8. Condensation produced by mixing.

path, but only on the initial and final states, we may consider the whole process as consisting (1) of a mixture giving supersaturated air as represented by P, followed by (2) condensation of water until the saturation vapor pressure is reached. The latter is the same type of isenthalpic process studied in Section 3, and can be represented on the diagram by the corresponding straight line PP'. The latent heat released is responsible for the heating from T to T'.

It may be noticed that the amount of liquid water produced per unit mass of air is given by the difference $q - q'_w \cong r - r'_w \cong (\varepsilon/p)(e - e'_w)$, where $(e - e'_w)$ is represented on the diagram by the segment $P'Q$. The corresponding concentration of liquid water per unit volume (cf. Section 2, Equation (20)) will be given by $\Delta c = (e - e'_w)/R_v T$.

This mechanism can in principle produce fogs (*mixing fogs*), but the amounts of condensed phase produced by it in the atmosphere are very low. It is however efficient in assisting other mechanisms (such as radiative cooling) to produce fogs.

A particular kind of mixing fog is *steam fog*, produced when cold air moves over warmer water. The layer in contact with the water will be saturated at the water temperature. As the cold air moves over this layer, a mixing process takes place, resulting in condensation. Due to the temperature stratification, these fogs are unstable, vertical stirring dissipating them into the drier air above. An example of this type of fog is the so-called 'Arctic Sea smoke', produced by very cold air passing over open water, e.g., in the North Atlantic or Baffin Bay or in open leads in the Arctic Ocean.

A further important example of adiabatic isobaric mixing processes occurs during the combustion of fuels when the exhaust gases are mixed with the atmosphere. When temperatures are low, fog may result from natural sources of combustion products and at upper levels condensation trails may be produced by aircraft. Since such effects are very important operationally, it is of interest to derive criteria for the occurrence of such phenomena. We may assume that condensation will be initiated, regardless of the temperature, only if the air in a mixed volume becomes saturated with respect to liquid water. At those temperatures for which this generally occurs the freezing nuclei content is usually adequate to cause transformation of the condensed phase to ice crystals, whose growth will continue until the vapor content is reduced to the ice-saturation value. The significance of the phenomenon will depend on the concentration of the condensed (ice) phase. For combustion processes at the surface, this concentration will depend on many factors – the rate of fuel consumption, the local topography, the intensity of vertical mixing, the horizontal wind speed, etc. In this case, a quantitative assessment of horizontal visibility (or of ice crystal concentration) is in general impossible. In the case of aircraft condensation trails, the situation is rather different since the generating element (the aircraft) is moving very rapidly relative to the air so that local accumulation of condensed phase is impossible. Accordingly, the aircraft problem will be analyzed in some detail, but it will be indicated how the primary criterion, for initiation of condensation, may be applied to surface combustion processes. We shall consider only a jet aircraft, for which all the heat produced by combustion is ejected into the exhaust gases.

We may imagine that the aircraft ejects heat and water vapor into a long thin cylinder along its flight path. The only significant diffusion and mixing will take place at right angles to this line so that one can visualize the resulting trail (visible or not) as consisting of a conical-shaped volume of air to the rear of the aircraft. The air along the central axis of this cone will most closely resemble the air in the original cyclindrical tube, and the air at the exterior limits will be essentially identical with undisturbed environmental air. We may assume that the heat and water vapor are diffused outwards in a similar manner and that radiative cooling plays a negligible role. For simplicity in numerical analysis, and because the geometry of the diffusion processes is not at all critical, we may imagine all sections normal to the cone axis to be completely mixed, over a cross section, A.

Let us first derive the conditions under which saturation with respect to liquid (or supercooled) water just occurs at a single value of A. Let us imagine that m_1 kg dry air plus $m_1 r$ kg water vapor are drawn into the jet engine, combined with F kg fuel (the fuel consumption per unit length of path) and exploded, adding FQ J of heat energy (Q is the heat of combustion, in J kg^{-1}, expressed in terms of gaseous products of combustion) and Fw kg water vapor to a volume A, containing originally m_2 kg dry air and $m_2 r$ kg water vapor.

Within the volume A, corresponding to a unit length of trail, the temperature will be in excess of that of the environment and will be denoted by $T + \Delta T$; the air will be assumed to be saturated with respect to liquid water and to contain a concentration E of the condensed phase (here assumed to be liquid). The First Principle of Thermodynamics enables us to state (neglecting the heat capacities of liquid water and combustion products other than water vapor; cf. Chapter IV, Equation (104), and notice that the heat term FQ must also be added in the present case)

$$FQ + l_v E A = (m_1 + m_2) \Delta T (c_{p_d} + r c_{p_v}). \tag{40}$$

The conservation of total water substance (r_t) requires that

$$(m_1 + m_2) r + Fw = (m_1 + m_2) r_w + EA = (m_1 + m_2) r_t. \tag{41}$$

The equation of state for the resulting gas phase is (cf. Chapter IV, Equation (70))

$$v = \frac{A}{(m_1 + m_2)(1 + r_w)} = \frac{R(T + \Delta T)}{p} = \frac{R_d(T + \Delta T)}{p}\left[1 + \left(\frac{1}{\varepsilon} - 1\right)r_w\right]. \tag{42}$$

Therefore

$$\frac{A}{m_1 + m_2} = \frac{R_d(T + \Delta T)}{p}\left(1 + \frac{r_w}{\varepsilon}\right). \tag{43}$$

Dividing Equation (40) by $(m_1 + m_2)$ and using Equation (43) gives

$$\Delta T\left[c_{pd} - \frac{R_d}{Ap}(FQ + l_v EA)\right] = \frac{R_d T}{Ap}(FQ + l_v EA), \tag{44}$$

where we have ignored some trivial terms in r and r_w. Making the additional yet equivalent assumption that vapor pressures are small relative to the total pressure, a similar treatment of Equation (41) gives, making appropriate substitutions for r and r_w,

$$\frac{\varepsilon e_w}{p} U_w = \frac{\varepsilon(e_w + \Delta e_w)}{p} + \frac{R_d(T + \Delta T)}{Ap}(EA - Fw). \tag{45}$$

If ΔT at the point in the trail with maximum of trail density E is small relative to T, we can invoke the Clausius-Clapeyron equation in the form

$$\Delta e_w = \frac{\varepsilon l_v e_w}{R_d T^2}\Delta T. \tag{46}$$

Substituting Equations (44) and (46) into Equation (45), and collecting terms, gives

$$E\left[\frac{\varepsilon l_v e_w}{p}\left(\frac{\varepsilon l_v}{R_d T} + U_w - 1\right) + c_{pd} T\right] + \frac{F}{A}\left[\frac{\varepsilon e_w Q}{p}\left(\frac{\varepsilon l_v}{R_d T} + U_w - 1\right) - c_{pd} Tw\right]$$

$$+ \frac{c_{pd}}{R_d}\varepsilon e_w(1 - U_w) = 0. \tag{47}$$

The coefficient of E will always be positive (since $\varepsilon l_v \gg R_d T$), and the final term zero (for saturated air) or positive. The coefficient of F/A can be either positive or negative. If positive, E will be everywhere negative (trail everywhere unsaturated). If the coefficient of F/A is negative, the trail density will be positive, at least for small A (and a maximum for vanishing A). If the environmental air is not saturated, the trail density becomes zero for some value of A, and whether or not a trail forms will depend on conditions close behind the aircraft (where ΔT and Δe_w may well be large). If the environmental air is saturated with respect to the liquid phase, a single criterion for positive trail density exists for all values of A (hence is valid for large A and small Δe_w and small ΔT, for which the above derivation is reasonably accurate). Thus, for saturated air, condensation can occur if

$$c_{pd} Tw > \frac{\varepsilon^2 e_w l_v Q}{R_d p T} \tag{48}$$

or, at a fixed temperature, if the pressure exceeds a critical pressure

$$p_c = \frac{\varepsilon^2 e_w l_v}{c_{p_d} R_d T^2} \frac{Q}{w}. \tag{49}$$

Inserting reasonable values for the ratio Q/w (appropriate for kerosene or a similar fuel) reveals that trails can only form at very low temperatures, achieved in the Arctic in the winter at the ground and elsewhere in the upper troposphere when the tropopause is high and cold. Since the air under these conditions would seldom, if ever, be saturated with respect to liquid water, it is necessary to carry out a more elaborate analysis than that given above if universal criteria for condensation trails are desired.

Even if temperatures in the trail, near the aircraft, are of the order of 10 to 20°C one can still assume $(e_w + \Delta e_w) \ll p$ but one can no longer treat e_w as a linear function of temperature. Let us, therefore, replace Equation (46) by a higher-order approximation. It is adequate to treat $l_v T^{-2}$ as virtually constant, so that

$$\Delta e_w \cong \frac{\varepsilon l_v e_w}{R_d T^2} \Delta T + \frac{e_w}{2} \left(\frac{\varepsilon l_v \Delta T}{R_d T^2} \right)^2. \tag{50}$$

Correct to this order of accuracy, one can use in the final term in Equation (47) the approximate relation, obtained by neglecting ΔT in Equation (43) when substituting in Equation (40),

$$(\Delta T)^2 \cong \frac{R_d T}{A p c_{p_d}} (FQ + l_v EA) \Delta T. \tag{51}$$

Substituting Equations (44), (50) and (51) into Equation (45), and collecting terms, gives

$$EB_1 + \frac{F}{A} B_2 + \left(\frac{F}{A}\right)^2 B_3 + B_4 = 0, \tag{52}$$

where B_1, B_2 and B_4 have the same values as they did in Equation (47), and where

$$B_3 = \frac{\varepsilon^3 e_w l_v^2 Q^2}{2 c_{p_d} R_d p^2 T^2}. \tag{53}$$

In the above derivation, the correction term has been obtained by maintaining only the first (Q^2) term in the expansion of $(FQ + l_v EA)^2$, since we are primarily interested in solutions for which E is very small and A is relatively small.

E will have a maximum value, E_m, for a value A_m of the trail cross-section such that $\partial E/\partial A$ is zero, yielding

$$\frac{F}{A_m} = -\frac{B_2}{2B_3} \tag{54}$$

$$E_m = \frac{B_2^2}{4B_1B_3} - \frac{B_4}{B_1}. \tag{55}$$

The critical conditions for a trail are those for which E_m is zero, or

$$B_2 + 2(B_3B_4)^{1/2} = 0. \tag{56}$$

Substituting for the values of the B parameters, the critical pressure, p_c, becomes

$$p_c = \frac{\varepsilon^2 e_w l_v}{c_{p_d} R_d T^2} \frac{Q}{w} \left\{ 1 + [2(1-U_w)]^{1/2} - \frac{R_d T}{\varepsilon l_v}(1-U_w) \right\}. \tag{57}$$

The final term inside the brackets can be ignored since $R_d T \ll \varepsilon l_v$. In can be seen that Equations (57) and (49) are identical for $U_w = 100\%$, but differ substantially if the air is relatively dry. This does not, however, imply a significant shift relative to temperature since the critical curves are steep when plotted on a $(T, \ln p)$ diagram. This can be demonstrated by logarithmic differentiation of Equation (57), assuming as before that e_w varies much more rapidly with temperature than does $l_v T^{-2}$. It follows (using the Clausius-Clapeyron equation) that

$$\left(\frac{\partial \ln p_c}{\partial T} \right)_{U_w} \simeq \frac{\varepsilon l_v}{R_d T^2}, \quad \text{or} \quad \left(\frac{\partial \ln p_c}{\partial \ln T} \right)_{U_w} \ll 1. \tag{58}$$

Condensation trails will occur for $p > p_c(T)$ or for $T < T_c(p)$, as defined by Equation (57).

This situation is illustrated on Figure VII-9, in terms of a $(T, \ln p)$ diagram. Thus, in zone III, condensation trails are impossible since supersaturation relative to liquid water is not observed in the atmosphere. In zone I condensation trails will always

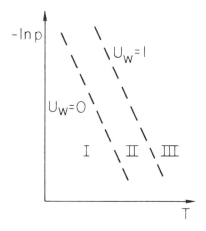

Fig. VII-9. Zones of an emagram relative to condensation trails.

form, regardless of the relative humidity. In zone II the formation of condensation trails, or their absence, will depend on the ambient relative humidity; formation will be possible if the (p, T) point lies to the left of the U_w line corresponding to the actual relative humidity.

Equation (57) can also be used to predict the formation of low temperature fog. It does not depend on the rate of fuel consumption but only on the ratio of heat to water vapor in the exhaust gases. For engines and furnaces of the normal type Q must be reduced by the amount abstracted for heating purposes or for the performance of work. Eventually, of course, such heat would reach the atmosphere (conduction through walls, frictional losses, etc.) so that if one does not deal with localized sources (a chimney or exhaust pipe) the full value of Q may be employed.

It is now important, at least for jet aircraft trails, to deduce the criteria for visible trails. It has been estimated that for a condensation trail to be visible from a distance the trail density (of ice phase), I in kg m^{-3}, must exceed 4×10^{-6} kg m^{-3}. This value will be taken as the critical trail density, I_c, and we shall investigate the conditions under which this value is just achieved at a single value of the cross-sectional area, A_c. This will correspond to a very small just-visible trail volume, moving along at a fixed distance behind the aircraft. For denser, longer and hence more-persistent trails, values of $I > I_c$ can be investigated. Fortunately, at a fixed p and r, temperatures only a few degrees colder than that yielding $I = I_c$ give a trail density very much greater than I_c.

We must now reformulate our basic equations to take into account the condensation, as ice particles, of I kg m^{-3}, and the resultant saturation of the air with respect to ice (provided of course that saturation relative to the liquid phase was achieved first). The equations are, of course, very similar to the previous set and can be written down immediately by analogy, viz.

$$IC_1 + \frac{F}{A} C_2 + \left(\frac{F}{A}\right)^2 C_3 + C_4 = 0, \tag{59}$$

where

$$C_1 = \frac{\varepsilon l_s e_i}{p} \left(\frac{\varepsilon l_s}{R_d T} + U_i - 1\right) + c_{p_d} T,$$

$$C_2 = \frac{\varepsilon e_i Q}{p} \left(\frac{\varepsilon l_s}{R_d T} + U_i - 1\right) - c_{p_d} Tw, \tag{60}$$

$$C_3 = \frac{\varepsilon^3 e_i l_s^2 Q^2}{2 c_{p_d} R_d p^2 T^2} \quad \text{and} \quad C_4 = \frac{c_{p_d}}{R_d} \varepsilon e_i (1 - U_i).$$

By analogy to Equation (55), it follows that a visible trail is just possible if

$$I_m = I_c = \frac{C_2^2}{4 C_1 C_3} - \frac{C_4}{C_1}. \tag{61}$$

Substituting from the set in Equation (60), and recalling that $\varepsilon l_s \gg R_d T$, one obtains

$$1 - U_i = \frac{1}{2}\left(1 - \frac{c_{p_d} R_d p T^2}{\varepsilon^2 e_i l_s} \frac{w}{Q}\right)^2 - I_c\left(\frac{\varepsilon l_s^2}{c_{p_d} p T} + \frac{R_d T}{\varepsilon e_i}\right). \tag{62}$$

For consistency with criterion (57), which must also be obeyed, we can introduce U_w by

$$U_i e_i = U_w e_w. \tag{63}$$

The final criterion curves can be obtained by compounding the two critical curves (just saturation re liquid water and $I_{max} = I_c$) for a given U_w value, choosing the segment at any pressure giving the lower critical temperature. The above equation is best solved by computing U_w as a function of T (or p) for a fixed pressure (or temperature) and interpolating the critical temperature (or pressure) for chosen U_w values.

With the above results, we may also investigate the criterion for intensification or dissolution of cirrus cloud when a jet aircraft flies through such cloud, or the parallel case of using a burner to attempt to dissipate fog when surface temperatures are low. In this case $I_c = 0$ and $U_i = 100\%$, so that the critical pressure below which cirrus consumption will occur is, from Equation (62),

$$p'_c = \frac{\varepsilon^2 e_i l_s}{c_{p_d} R_d T^2} \frac{Q}{w}. \tag{64}$$

It may be noted that the converse effect (i.e., condensation in the ice phase) is often responsible for dense ice crystal fog at very low temperatures due to aircraft warm-up and take-off, at times sufficiently severe to cause a temporary cessation of flying.

7.6. Adiabatic Expansion in the Atmosphere

The processes of adiabatic expansion (or compression) are particularly important, because they describe the transformations taking place when an air mass rises (or descends) in the atmosphere. They are therefore a part of the study of convection, and we may consider what happens when a parcel of the atmosphere rises without mixing with its environment, that is, during the adiabatic expansion of moist air. In the more general case, the entire atmosphere is assumed to rise; the relations are, of course, identical.

The first stage will be a moist adiabatic expansion of only one gaseous phase; this is a simple process which was already considered in Chapter VI, Section 4, where we saw that it differs very little from a dry adiabatic expansion. In other words, we can write with good approximation $\varkappa \cong i_d$ (cf. Chapter IV, Equation (89)), and represent the process by the equation

$$\theta = T\left(\frac{1000}{p}\right)^{\varkappa_d} \tag{65}$$

(p in mb), which on a diagram corresponds to the dry adiabats; each curve is characterized by a potential temperature θ, which must be the value of T where the curve intersects the 1000 mb isobar. Along each of these curves, entropy has a constant value $s = c_{p_d} \ln \theta$ + const. These relations are shown in Figure VII-10 for a tephigram.

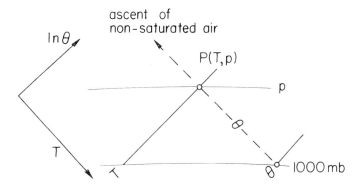

Fig. VII-10. Non-saturated adiabatic expansion on a tephigram.

The decrease in temperature will continue until saturation is reached, that is, until the temperature is such that the air humidity is that of saturation (r_w becomes equal to r). From that moment on, the second stage proceeds. In this stage two possibilities may be considered. The water (or ice) produced by condensation may remain in the air (cloud), so that if we reverse the process, the same water will evaporate during the adiabatic compression along the descent; in this case, we shall call it a *reversible saturated adiabatic expansion*. It is implied that the process must be slow enough to be considered reversible. Being adiabatic and reversible, it is also isentropic.

We may also assume that the water or ice falls out of the system as soon as it is produced. In this case we are dealing with an open system, and the process is called *pseudo-adiabatic*.

Often an intermediate process will operate in the atmosphere, where part (not all) of the water or ice condensed during the ascent falls out as precipitation.

Due to the condensation heat, cooling is slower in a saturated than in a moist expansion. But as condensation must proceed along with the cooling (r_w decreases), saturated adiabats must have, as was explained in Chapter VI, Section 4, an intermediate slope between moist adiabats and the saturation mixing ratio isopleths.

It has been customary to talk of a *rain stage*, a *snow stage*, and a *hail stage*. By the rain stage is meant the reversible adiabatic expansion during the ascent of air, with production of liquid water; the snow stage refers to the production of ice. Of course the temperature (and therefore the height in the atmosphere) determines whether we are in one or the other stage. The transition does not occur in general at 0 °C, but at lower temperatures. At air temperatures less than -40 °C, only sublimation occurs, because at those low temperatures liquid droplets freeze spontaneously, so that only ice clouds can exist. Mixed clouds can exist at intermediate temperatures between

0 °C and −40 °C, but obviously they are not in thermodynamic equilibrium; the ascent of the air will produce both condensation on droplets and sublimation on ice crystals, while a continuous process of distillation from the droplets (with higher vapor pressure) to the ice crystals (with lower vapor pressure) must occur. The so-called hail stage assumed that at the level with temperature 0 °C, the cloud water would freeze isentropically (reversible adiabatic freezing). Nothing of that kind ever occurs in the atmosphere, where freezing of droplets occurs only (at temperatures in excess of −40 °C) on *ice nuclei* (particles that favor freezing), and these only start being active at temperatures considerably below 0 °C, in increasing numbers as the temperature decreases; in these conditions the freezing is of course irreversible. Our analysis (in Section 10) will deal with the type of hail stage that actually occurs.

We shall first consider the attainment of saturation by adiabatic expansion of moist air (Section 7), and in later sections (Sections 8 and 9) the saturated expansion.

7.7. Saturation of Air by Adiabatic Ascent

In Section 1 we have dealt with the saturation by isobaric cooling. We shall now consider saturation due to adiabatic expansion by ascent in the atmosphere.

If we differentiate logarithmically the definition of U_w, we have

$$d \ln U_w = d \ln e - d \ln e_w. \tag{66}$$

During the ascent $e/p = N_v$ is a constant. Therefore, from Poisson's equation

$$Tp^{-\varkappa} = \text{const.} \tag{67}$$

we derive

$$Te^{-\varkappa} = N_v^{-\varkappa} \text{const.} = \text{const.}'. \tag{68}$$

That is, the partial pressure also obeys Poisson's equation. Differentiating logarithmically:

$$d \ln T = \varkappa \, d \ln e. \tag{69}$$

Introducing this expression and Clausius-Clapeyron's equation, we obtain for the logarithmic variation of relative humidity:

$$d \ln U_w = \frac{1}{\varkappa} d \ln T - \frac{l_v}{R_v T^2} dT. \tag{70}$$

Here the first term on the right hand side gives the change due to the decrease in pressure p (and therefore in e), while the second term measures the influence of the decrease in temperature and therefore in e_w and is of the same form as for an isobaric cooling (cf. Section 1). The two terms have opposite signs, so that in principle an adiabatic expansion could lead to a decrease as well as to an increase in U_w.

Equation (70) can also be written:

$$\frac{dU_w}{dT} = \frac{U_w}{T}\left(\frac{1}{\varkappa} - \frac{l_v}{R_v T}\right) = \frac{U_w}{T}\left(\frac{c_p T - \varepsilon l_v}{R_d T}\right). \tag{71}$$

The expression within brackets is <0 if

$$T < \varepsilon l_v / c_p \cong 1500 \,\text{K} \tag{72}$$

a condition which always holds in the atmosphere. Therefore U_w increases as T decreases, for an adiabatic expansion.

If we represent the initial state of the ascending air on a vapor pressure diagram by the image point P (see Figure VII-11), the arrow starting from P indicates the adiabatic ascent described by Equation (69). As the air temperature T changes, the saturation vapor pressure will simultaneously change according to the arrow starting from S, along the saturation curve, which is described by the Clausius-Clapeyron equation.

If we integrate Equation (70), we obtain

$$\ln \frac{U_w}{U_{w0}} = \frac{1}{\varkappa} \ln \frac{T}{T_0} + \frac{l_v}{R_v}\left(\frac{1}{T} - \frac{1}{T_0}\right), \tag{73}$$

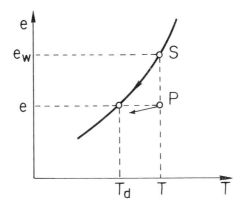

Fig. VII-11. Non-saturated adiabatic expansion on a vapor pressure-temperature diagram.

where T_0, U_{w0} refer to initial conditions. For $U_w = 1$

$$-\ln U_{w0} = \frac{1}{\varkappa} \ln \frac{T_s}{T_0} + \frac{l_v}{R_v}\left(\frac{1}{T_s} - \frac{1}{T_0}\right) \tag{74}$$

which can be solved numerically for the *saturation temperature* T_s. In the atmosphere the ascent can always lead to saturation, i.e., T_s can always be reached. On a tephigram, this means that starting from the image point and rising along the dry adiabat, we shall eventually reach the vapor line corresponding to the value r of the air (see Figure VII-12). On the vapor pressure diagram (Figure VII-11) it means that the adiabat starting from P will reach the saturation curve.

We shall see later that, along a dry adiabat, the temperature drops approximately 10°C for every km. We want now to make an estimation of the height at which

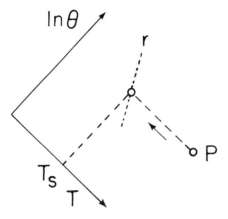

Fig. VII-12. Saturation by adiabatic expansion, on a tephigram.

saturation will be attained. For that purpose we shall first inquire what is the variation of the dew point temperature along the adiabat. This is given by the Clausius-Clapeyron equation (cf., Figure VII-11):

$$dT_d = \frac{R_v T_d^2}{l_v} d \ln e. \tag{75}$$

Introducing Equation (69):

$$dT_d \cong \frac{R_v T_d^2}{\varkappa l_v} \frac{dT}{T} = \frac{c_p T_d^2}{\varepsilon l_v} \frac{dT}{T}. \tag{76}$$

Writing $T_d \sim T \sim 273 \text{ K}$, and using finite differences, we obtain the approximate relation

$$\Delta T_d \cong \tfrac{1}{6} \Delta T. \tag{77}$$

That is, T_d decreases approximately one sixth of the temperature drop, along an adiabatic ascent. Figure VII-13 shows this result on the vapor pressure diagram. The same relations are shown in Figure VII-14 on a tephigram. In this last representation, if we consider that r remains constant during the unsaturated ascent, it is obvious that T_d will slide along an equisaturated line ($r = $ const.).

Let us now consider the variations of T and T_d during ascent from any level z_0 to the saturation level $z_s = z_0 + \Delta z$. At the saturation level, T becomes equal to T_d and to T_s. Taking into account the lapse rate of T along an adiabat, and the derived relation for ΔT_d, we may write:

$$T - T_s \cong 10 \, \Delta z \quad (\Delta z \text{ in km})$$

$$T_d - T_s \cong \tfrac{1}{6}(10 \, \Delta z).$$

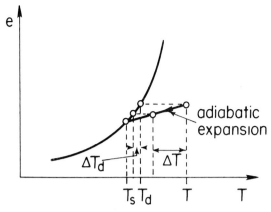

Fig. VII-13. Variation of dew-point and temperature during an adiabatic expansion, on a vapor pressure-temperature diagram.

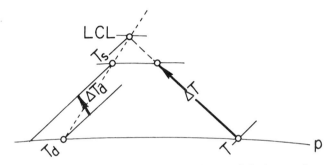

Fig. VII-14. Variation of dew-point and temperature during an adiabatic expansion, on a tephigram.

Subtracting:

$$T - T_d \cong \tfrac{5.0}{6} \Delta z$$

$$\Delta z \cong 0.12(T - T_d)\,\text{km} \cong 400(T - T_d)\,\text{ft}. \tag{78}$$

This relation allows an approximate estimate of the height of the condensation level for an adiabatic ascent. This will give the base of cumuli, provided that these are actually formed by air rising from z_0.

The level at which an air mass attains saturation by adiabatic ascent is called the *lifting condensation level* (usually abridged LCL), and the corresponding temperature is the saturation temperature T_s.

7.8. Reversible Saturated Adiabatic Process

Our system will be a parcel of cloud which rises, expanding adiabatically and reversibly; it will be a closed system, keeping all the condensed water. Being adiabatic and reversible, the process is isentropic.

The same derivation and final formula will also be valid for an ice cloud, with the appropriate substitutions (sublimation latent heat for vaporization latent heat, specific heat capacity of ice for specific heat capacity of water, etc.).

We have already calculated the entropy expression for this system (Chapter IV, Equation (119)). It is:

$$S = (m_d c_{p_d} + m_t c_w)\ln T - m_d R_d \ln p_d + \frac{m_v l_v(T)}{T} + \text{const.} \qquad (79)$$

Dividing by m_d and considering that the entropy must remain constant:

$$(c_{p_d} + r_{t,w} c_w)\ln T - R_d \ln p_d + \frac{r_w l_v}{T} = \text{const.} \qquad (80)$$

where $r_{t,w} = m_t/m_d = r_w + m_w/m_d$, $p_d = p - e_w$; the subscript w indicates saturation values.

In differential form, Equation (80) becomes:

$$(c_{p_d} + r_{t,w} c_w)\, d\ln T - R_d\, d\ln p_d + d\left(\frac{r_w l_v}{T}\right) = 0. \qquad (81)$$

Equation (80), or (81), describes the reversible saturated adiabatic process. As we have seen before, our system is bivariant. If T and p_d are considered as the independent variables, r_w can be expressed (Chapter IV, Equation (79)) as a function of e_w: $r_w = \varepsilon e_w/p_d$; and e_w, in its turn, as a function of T, through Clausius-Clapeyron's integrated equation (Chapter IV, Section 8). Equation (80) gives thus a relation between the two independent variables (whichever we choose) and determines a curve in the plane T, p_d or in any usual diagram.

If in Equation (81) we neglect $r_{t,w} c_w$ as compared with c_{p_d} and e_w as compared with p_d, we obtain the approximate formula*

$$c_{p_d} \ln T - R_d \ln p + \frac{r_w l_v}{T} = \text{const.} \qquad (82)$$

and considering that l_v varies slowly with T, we obtain the approximate differential formula

$$c_{p_d}\, d\ln T - R_d\, d\ln p + l_v\, d\left(\frac{r_w}{T}\right) = 0. \qquad (83)$$

7.9. Pseudoadiabatic Process

Equations (80) and (81) of Section 8 depend upon the value of r_t. This is a different constant for each different system considered, varying with the concentration of liquid

---

* For high temperatures, this becomes a rough approximation. If $r_{t,w}$ reaches a value of 0.05, $r_{t,w} c_w$ becomes equal to about 20% of c_{p_d}, and it can no longer be neglected.

water. For a given point T, p in a diagram, the vapor saturation mixing ratio r_w is also determined, but the liquid water mixing ratio m_w/m_d is an arbitrary parameter of the system. If we are considering rising air (expanding adiabatically), its value will depend on the level at which the air first became saturated. Therefore, through each point of a diagram an infinite number of reversible saturated adiabats will pass, differing (very little) from each other by the term depending on the liquid water content.

In order to avoid this inconvenience, aerological diagrams have recourse to another process of saturated expansion which gives uniquely defined curves. This is called the *pseudoadiabatic process*, and it is assumed in it that all the condensed water (or ice) falls out of the system as soon as it is produced. The system is an open one. However, we may easily find the relation in which we are interested, starting from the formulas of Section 8.

We must have at any instant

$$m_w = 0; \quad r_{t,w} = r_w.$$

We may consider the condensation and elimination of a mass dm_w of water as occurring in two stages: (1) reversible saturated adiabatic expansion, with condensation of a mass dm_w of water, and (2) the water leaving the system. In the first stage we may apply Equation (81), substituting r_w for $r_{t,w}$:

$$(c_{p_d} + r_w c_w) \, d \ln T - R_d \, d \ln p_d + d\left(\frac{r_w l_v}{T}\right) = 0. \tag{84}$$

In the second stage, the entropy of the system will decrease by $-s_w \, dm_w$, but this process will not affect the values of T and p. As for $r_{t,w}$, it will decrease by $-dm_w/m_d$, so that it will remain equal to $r_w(T)$. Therefore, Equation (84) describes the variation of T and p in a pseudoadiabatic process. If we want the integrated form, we shall have to remember that r_w depends now on T (which $r_{t,w}$ did not):

$$c_{p_d} \ln T + c_w \int r_w \, d \ln T - R_d \ln p_d + \frac{r_w l_v}{T} = \text{const.} \tag{85}$$

Obviously, in the reverse process (pseudoadiabatic saturated compression), one has to assume that water (or ice) is being introduced from outside at the instantaneous temperature of the system and at the necessary rate to maintain it exactly saturated.

If we now make the same approximations as in Section 8, we obtain again for pseudoadiabatic processes Equation (83), which is usually applied when precision is not required. Therefore, in approximate calculations no difference is made between the two processes. In aerological diagrams, the saturated adiabats correspond as explained to the pseudoadiabatic process. The cooling is slightly greater in pseudoadiabatic than in reversible expansion (for the same pressure change).

7.10. Effect of Freezing in a Cloud

We shall consider now the effect of freezing, assuming that this occurs at a given level during the air ascent, i.e., at constant pressure. We shall also assume that we are dealing with an adiabatic, and therefore isenthalpic process. We can still call this the 'hail stage', but the initial temperature can have any value between 0 °C and −40 °C; it is an irreversible process.

As the water freezes, latent heat is released. Also, the air initially saturated with respect to water at the initial temperature T, will be supersaturated with respect to the frozen droplets; sublimation will occur, releasing an additional amount of latent heat, until the saturation vapor pressure with respect to ice is reached, at the final temperature T'.

As the enthalpy change only depends on the initial and final states, we can calculate it as if the total process occurred in three steps:

(1) Water freezes at constant temperature T.

(2) Vapor condenses on the ice at constant temperature T, until the water vapor pressure reaches the saturation value over ice at T'.

(3) The whole system is heated from T to T'.*

This can be schematically summarized as follows:

$$\text{condensed phase:} \quad \begin{Bmatrix} \text{water} \\ T \\ e_w(T) \end{Bmatrix} \xrightarrow{(1)} \begin{Bmatrix} \text{ice} \\ T \\ e_w(T) \end{Bmatrix} \xrightarrow{(2)} \begin{Bmatrix} \text{ice} \\ T \\ e_i(T') \end{Bmatrix} \xrightarrow{(3)} \begin{Bmatrix} \text{ice} \\ T' \\ e_i(T') \end{Bmatrix}$$

The sum of the enthalpy changes for the three steps is the total change, and this must be zero.

We may consider for convenience a system containing unit mass of dry air, an amount r_L of liquid water ($r_L = m_w/m_d$) and the saturation mixing ratio of water vapor at the initial temperature $r_w(T)$. The three changes in enthalpy will be:

$$\Delta H_1 = -l_f r_L \tag{86}$$

$$\Delta H_2 = -l_s [r_w(T) - r_i(T')] \tag{87}$$

$$\Delta H_3 = [c_{p_d} + r_i(T')c_{p_v} + r_S c_i](T' - T)$$

$$= c_p(T' - T) \tag{88}$$

where $r_i(T')$ is the saturation mixing ratio over ice at the final temperature T', $r_S = r_L + r_w(T) - r_i(T')$ is the final mass of ice, and we have called c_p the average specific heat capacity of the final system.

---

* It can be imagined that the latent heat released in (1) is given to an external source, and that in (3) the necessary heat is again received from an external source.

We can express the mixing ratios in Equation (87) in terms of vapor pressures (formula in Chapter IV, Equation (82)):

$$r_w(T) \cong \frac{\varepsilon e_w(T)}{p} \qquad (89)$$

$$r_i(T') \cong \frac{\varepsilon e_i(T')}{p} \qquad (90)$$

and use the Clausius-Clapeyron equation to express e_i as a function of the initial temperature, assuming that $(T'-T)$ is small enough to be treated as a differential:

$$e_i(T') = e_i(T) + \frac{l_s e_i(T)}{R_v T^2}(T' - T). \qquad (91)$$

Introducing Equations (89), (90), (91) into (87), we have

$$\Delta H_2 = -\frac{\varepsilon l_s}{p}\left[e_w(T) - e_i(T) - \frac{l_s e_i(T)}{R_v T^2}(T' - T)\right]$$

$$= -r_w(T)l_s\left[1 - \frac{e_i(T)}{e_w(T)}\right] + \frac{r_i(T)l_s^2}{R_v T^2}(T' - T). \qquad (92)$$

Introducing Equations (86), (88) and (92) in

$$\Delta H_1 + \Delta H_2 + \Delta H_3 = 0 \qquad (93)$$

and solving for $(T'-T) = \Delta T$, we finally obtain

$$\Delta T = \frac{l_f r_L + l_s r_w\left(1 - \dfrac{e_i}{e_w}\right)}{c_p + \dfrac{r_i l_s^2}{R_v T^2}}. \qquad (94)$$

Here the saturation values are taken at T. The term c_p contains $r_i(T')$, which is not known if we are calculating ΔT; however $r_i(T')c_{p_v} \ll c_{p_d}$ (cf. Equation (88)) and $r_i(T)$ can be used for $r_i(T')$. If needed, successive approximations would rapidly improve the computation.

Formula (94) gives the increase in temperature due to the freezing of the cloud water. For individual cloud parcels, this increase adds to the buoyancy (upward thrust due to the difference of density with the environment: cf., Chapter IX, Section 2) which is important in the dynamics of such a cloud. In reality, however, freezing does not occur suddenly at one level, but gradually over a temperature interval. If this were to be taken into account, the problem should be integrated with that of the saturated expansion. The distribution of freezing along the ascent should then be known or assumed. The system would not be in equilibrium and the vapor pressure would have some intermediate value between those of saturation over water and saturation over ice. The solution of the problem (curve $T = f(p)$) would depend on the model by which the process might be approximated.

Especially at low temperatures, the warming accompanying freezing will depend critically on the liquid water content of the cloud. In the limiting case, when r_w and r_i become negligible,

$$\Delta T = \frac{r_L}{c_p} l_f. \tag{95}$$

At $-30\,°C$, if $r_L = 10 \text{ g kg}^{-1}$ (a high liquid water content), Equation (95) gives $\Delta T = 3.3\,K$. This warming could be a very significant factor in cloud development.

7.11. Polytropic Expansion

Vertical motions in the atmosphere can have velocities varying within a very wide range. Violent convective processes may entail updrafts of several tens of meters per second. On the other extreme, some synoptic situations are associated with very slow motions of horizontally extended layers; vertical velocities can be as small as 0.1 to 1 cm s^{-1}. In the latter extreme cases the expansion or compression cannot be considered as strictly adiabatic, as it has been assumed so far for vertical displacements; the exchange of radiant energy in 'long wave' (i.e. in the terrestrial and atmospheric range of wave lengths, from 4 to several tens of micrometers) with other horizontal layers, with the ground and with outer space cannot be ignored any more without introducing appreciable error. Such radiant exchanges produce temperature changes in the air of the order of 1–2 K day^{-1}. For the sake of comparison, the adiabatic cooling of air rising at 1 cm s$^{-1} \cong 0.86$ km day^{-1} would be 8.4 K day^{-1} (as it will be seen in Chapter IX, Section 3) and that of air rising at 0.1 cm s$^{-1} = 86$ m day^{-1} would be 0.84 K day^{-1}.

These non-adiabatic expansions or compressions can be approximated by the polytropic processes described in Chapter II, Section 8. In order to find the relation between the values of the coefficients k and n in the formulas (61) of that section and the conditions of these atmospheric processes, we can write the first principle in specific quantities as

$$\delta q = c_p\, dT - v\, dp, \tag{96}$$

differentiate

$$T = \text{const.}\, p^{(1-1/n)} = \text{const.}\, p^k \tag{97}$$

logarithmically, which gives

$$dT = k\frac{T}{p}\, dp \tag{98}$$

and make use of the hydrostatic equation

$$dp = -g\varrho\, dz \tag{99}$$

(where g = gravity, dz = air ascent), which will be studied in Chapter IX, Section 2. By eliminating dT and dp from the three equations (96), (98) and (99), it is easily found that

$$k = \varkappa\left(1 - \frac{1}{g}\frac{\delta g}{dz}\right) \tag{100}$$

and

$$n = \frac{1}{1-k} = \frac{1}{1 - \varkappa\left(1 - \frac{1}{g}\frac{\delta g}{dz}\right)}. \tag{101}$$

$\delta q/dz$ is the heat absorbed by the air per unit mass and unit length of ascent. If the information available is given in terms of rate of temperature charge λ due solely to radiation exchange

$$\lambda = \frac{dT}{dt} = \frac{1}{c_p}\frac{\delta q}{dt} \tag{102}$$

and velocity of ascent

$$U = \frac{dz}{dt} \tag{103}$$

we can write

$$\frac{\delta q}{dz} = \frac{c_p \lambda}{U}. \tag{104}$$

Thus, coming back to the figures mentioned before, if for instance $\lambda = \pm 2\,\text{K day}^{-1}$ (upper and lower signs corresponding to warming and cooling, respectively) and $U = 0.86\,\text{km day}^{-1}$, we obtain (assuming dry air), with $c_p = 1005\,\text{J kg}^{-1}\text{K}^{-1}$, $g = 9.81\,\text{m s}^{-2}$:

$$\frac{\delta q}{dz} = \pm 2.34\,\text{J kg}^{-1}\text{m}^{-1}$$

and $k = 0.219$ or 0.355, corresponding to the upper or the lower sign, respectively; this is to be compared with the adiabatic coefficient $\varkappa_d = 0.286$. The corresponding values of n will be 1.28 and 1.55, respectively, to be compared with $\eta_d = 1.40$.

7.12. Vertical Mixing

We shall consider now the mixing of air masses along the vertical (by turbulent and/or convective processes). The analysis becomes complicated in this case by the continuous variation of p, T and r with height. We shall consider first two isolated masses m_1 and m_2 at the pressure levels p_1 and p_2 (with temperatures T_1 and T_2), which move to another pressure p and mix. We have then a first stage of adiabatic expansion or compression, for which

$$T'_1 = T_1\left(\frac{p}{p_1}\right)^{\varkappa}; \quad T'_2 = T_2\left(\frac{p}{p_2}\right)^{\varkappa}. \tag{105}$$

The specific humidities q_1 and q_2 (or the mixing ratios r_1 and r_2) are preserved in this stage.

The second stage will be an isobaric adiabatic mixture, already considered in Section 4; formulae (35) (or the approximate Equation (37)), (36) and (39) will apply, as well as formulae similar to those of Equation (35) and (37) for the potential temperature; in particular Equation (38), where θ_1 and θ_2 are the initial values, which do not vary during the adiabatic expansion or compression. That is, the potential temperature of the mixture is approximately given by the weighted average of the potential temperatures of the two air masses. The third stage will consist in taking the two air masses to their original pressure levels. The location of the auxiliary level p used for the derivation is of course immaterial.*

Let us consider now an atmospheric layer with a thickness $\Delta p = p_2 - p_1$, and let us suppose that it mixes vertically. We can imagine that the process is performed by bringing the whole layer to the same level p, mixing it isobarically and redistributing it in the original interval Δp. The mixture will consist of air with a potential temperature θ equal to the weighted average for the whole layer. The vertical redistribution will preserve the value of θ, since it consists of adiabatic expansions or compressions. Therefore, when the layer is thoroughly mixed, θ will be constant with height.

The mass per unit area of an infinitesimal layer dz will be $dm = \varrho\, dz$, where ϱ is the density at the height of the layer. We shall see in Chapter VIII, Section 2 that (assuming hydrostatic equilibrium)

$$\varrho\, dz = -\frac{1}{g}\, dp, \tag{106}$$

where g is the acceleration of gravity and $-dp$ the variation of p in dz. Therefore

$$\bar{\theta} = \frac{\int_0^m \theta\, dm}{m} = \frac{\int_0^z \theta \varrho\, dz}{\int_0^z \varrho\, dz} = -\frac{\int_{p_1}^{p_2} \theta\, dp}{p_1 - p_2}. \tag{107}$$

Similar expressions would give the final values of q and r. The temperature T will have a distribution along p given by the adiabat for $\bar{\theta}$, i.e.,

$$T = \bar{\theta}\left(\frac{p}{1000}\right)^{\varkappa}. \tag{108}$$

---

* We might also have assumed that both masses move to $p = 1000$ mb, in which case $T_1' = \theta_1$ and $T_2' = \theta_2$.

This case is illustrated by Figure VII-15 on a tephigram, where the solid line shows the initial temperature distribution, the dashed line is the final adiabat $\bar{\theta}$,* and the dotted line (vapor line corresponding to the final uniform value of $\bar{r}$) shows that the final distribution does not reach the saturation level.

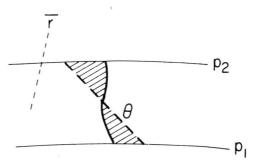

Fig. VII-15. Vertical mixing without condensation, on a tephigram.

If the vapor line $\bar{r}$ intersects the adiabat $\bar{\theta}$, from that level upwards condensation will occur and the final temperature distribution will follow the saturated adiabat. The intersection level is called the *mixing condensation level* (MCL). This case is illustrated by Figure VII-16.

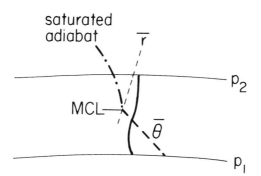

Fig. VII-16. Mixing condensation level, on a tephigram.

7.13. Pseudo- or Adiabatic Equivalent and Wet-Bulb Temperatures

Formula (29) is not convenient for a determination of T_{ie} and T_{iw} on aerological diagrams. But another two parameters, closely related to the above, may be defined, that can be easily found on the diagrams.

Let P be the image point of the air, with a mixing ratio r (see Figure VII-17). Let us

---

* Compensation of areas is not strictly applicable here, but it will give a good approximation if the thickness of the layer is not too large.

assume that the air expands adiabatically until it reaches saturation; we have seen in Section 7 that this is always possible. This will happen at the intersection of the dry adiabat with the vapor line r. This is the *characteristic point* of the air, and has been designated as P_c on the diagram. Its temperature is the saturation temperature T_s. A saturated adiabat must also pass through that point; if we now follow it towards increasing values of the pressure, we attain the point P_w at the original pressure p. The temperature of P_w is called the *adiabatic wet-bulb temperature* or the *pseudo-wet-*

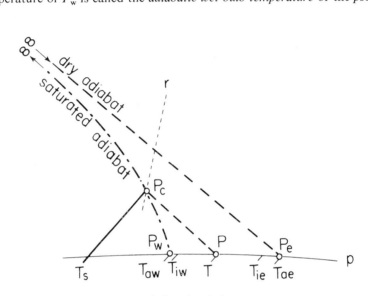

Fig. VII-17. Wet-bulb and equivalent temperatures.

bulb temperature. In order to be able to follow the curve $P_c P_w$, we must imagine that we are evaporating water into the air, so as to keep it saturated at increasing temperatures. According to the definition of these curves on the diagrams (pseudoadiabatic process, Section 9), we must consider that liquid water is being introduced in the precisely necessary amounts to maintain saturation, at the varying temperature of the air, and evaporated into it. If we compare this process with the experimental definition of T_{iw} (through the wet-bulb psychrometer) we find that while in the latter case all the added water was initially at the temperature T_{iw} (and its heat capacity was finally neglected in the analytical definition of T_{iw}), in the process now under consideration it was initially at varying temperatures between T_s and T_{aw} and after evaporation it had to be warmed to T_{aw}, subtracting heat from the air. For this reason T_{aw} is smaller than T_{iw}, but only by a small difference, usually not exceeding 0.5 °C.

Let us assume now that again we expand the air, starting from P. But this time we continue expanding after P_c. The air will then follow the saturated adiabat, the water vapor will condense, and we assume that it falls out of the system during the process. If we continue this process indefinitely, the curve will tend asymptotically to a dry adiabat, as the water vapor content becomes negligible (this is indicated in Figure VII-17 by ∞). Along this stage, water vapor cools together with the air before condensing and falling out at the variable temperature of condensation. Once all the vapor has thus been eliminated, we compress again the air to the original pressure p. In this last stage, the air is dry and will follow a dry adiabat, reaching the final temperature T_{ae} which we call the *adiabatic equivalent* or *pseudo-equivalent temperature*. As for T_{ie}, the change (not the process) undergone by the dry air component of the system is isobaric. But while for T_{ie} the condensed water remains at intermediate temperatures between T and T_{ie} (and its heat capacity is finally disregarded), in the process now under consideration it remains at varying temperatures below T_s; thus, $T_{ae} > T_{ie}$. The difference is in this case larger than for the wet-bulb temperatures, and cannot be neglected in general.

If in the case of the pseudo-wet-bulb temperature we extend the saturated adiabat to the 1000 mb isobar, the intersection determines, by definition, the *pseudo-wet-bulb potential temperature* θ_{aw}. Similarly, by extending the dry adiabat from T_{ae} to the intersection with the 1000 mb isobar, we determine the *pseudo-equivalent potential temperature* θ_{ae}.

7.14. Summary of Temperature and Humidity Parameters. Conservative Properties

We summarize now the different temperatures that have been defined, with reference to the sections where they were introduced.

T = temperature

Dry temperatures

    T_v = virtual temperature (Chapter IV, Section 11).

    $T_e = T_{ie}$ = (isobaric) equivalent temperature (Section 3).

    T_{ae} = adiabatic equivalent, or pseudo-equivalent, temperature (Section 13).

Saturation temperatures

    T_d = dew point temperature (Section 1).

    T_f = frost point temperature (Section 1).

    $T_w = T_{iw}$ = (isobaric) wet-bulb temperature (Section 3).

    T_{aw} = adiabatic wet-bulb, or pseudo-wet-bulb, temperature (Section 13).

    T_s = saturation temperature (Section 7).

Potential temperatures

    θ = potential temperature (Chapter II, Section 7).

    θ_v = virtual potential temperature (Chapter IV, Section 13).

    $\theta_e = \theta_{ie}$ = (isobaric) equivalent potential temperature.*

    θ_{ae} = adiabatic equivalent, or pseudo-equivalent, potential temperature (Section 13).

    $\theta_w = \theta_{iw}$ = (isobaric) wet-bulb potential temperature.*

    θ_{aw} = adiabatic wet-bulb, or pseudo-wet-bulb, potential temperature (Section 13).

* Defined in a similar way to the corresponding pseudo-potential temperature (see Figure VII-18).

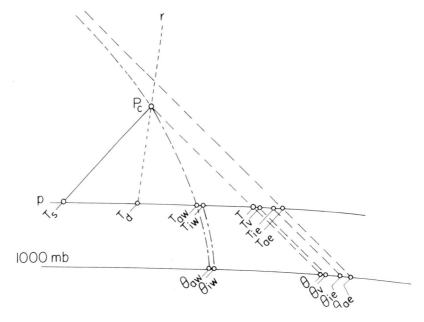

Fig. VII-18. Summary of temperature-humidity parameters.

Among the non-potential temperatures we have the following inequality relations:

$$T_s < T_d < T_{aw} < T_{iw} < T < T_v < T_{ie} < T_{ae}.$$

Figure VII-18 gives a summary of these parameters, as they are determined on a tephigram.

The reason why all these parameters are defined lies in their conservative properties regarding certain meteorological processes. Thus, for instance, an air mass becoming warmer or cooler, isobarically, maintains the same value of its dew point. An air mass rising or descending adiabatically, without condensation or evaporation, keeps constant its potential temperature; if evaporation or condensation takes place, θ varies but θ_{aw} and θ_{ae} preserve their values. A cool, moist, air mass that rises along the slopes of a mountain range, drying by precipitation, may eventually descend along the slopes on the lee side, arriving as a dry, hot air mass (Föhn); the air mass may be recognized to be the same as before if its potential temperatures θ_{aw} or θ_{ae} remain the same.

These parameters may thus be used to identify air masses undergoing a series of transformations.

Table VII-1 summarizes the conservative properties of several temperature and humidity parameters, with respect to the main processes referred to above.

TABLE VII-1

Conservative properties of several parameters, C = conservative; NC = non-conservative

Property	Process:			
	Isobaric warming or cooling (without condensation or vaporization)	Isobaric vaporization or condensation	Non-saturated adiabatic expansion	Saturated adiabatic expansion
U_w	NC	NC	NC	C
e or T_d	C	NC	NC	NC
q or r	C	NC	C	NC
T_{aw} or T_{ae}	NC	C	NC	NC
θ	NC	NC	C	NC
θ_{aw} or θ_{ae}	NC	C	C	C

PROBLEMS

1. An air mass has a temperature of 20 °C at 1000 mb pressure, with a mixing ratio of 10 g kg^{-1}. After a certain time, the same mass, without mixing with the environment has acquired a temperature of 10 °C and a pressure of 750 mb. Knowing that the dew-point is initially 14.0 °C, calculate analytically the initial and final values of the vapor pressure e and relative humidity U_w, and the final dew-point temperature T_{d_f}. Do not use tables of saturated vapor pressures.

2. During the formation of a radiation fog, 1 cal g^{-1} is lost after saturation started, at 10 °C. The pressure is 1000 mb. What is the final temperature? What was the decrease in vapor pressure, and what is the concentration of the fog, in grams of liquid water per cubic meter?

    Make an approximate calculation, treating the differences as differentials, and using the approximate value 12 mb for the saturation vapor pressure.

3. The isobaric equivalent temperature is defined as the temperature that the humid air would attain if all its water vapor were condensed out at constant pressure, and the latent heat released used to warm the air. Using the expression for the enthalpy of Chapter IV, Equation (104), show that a more accurate expression of the isobaric equivalent temperature than Equation (27) is given by the following formula

$$\ln \frac{l_v(T_{ie})}{l_v(T)} = \frac{c_{pv} - c_w}{c_{pv}} \ln\left(1 + \frac{c_{pv}}{c_{pd}} r\right)$$

4. An air parcel, initially at 10 °C and with a water vapor pressure of 3.5 mb, undergoes an isobaric, adiabatic wet-bulb process to saturation. Derive in approximate form (assuming the heat capacity of the water substance is negligible) an expression giving the slope of the line representing this process on a $e - T$ (vapor pressure – temperature) diagram. Draw this line on such a diagram for a total pressure of 1000 mb, and obtain the approximate values of the wet-bulb temperature and of the equivalent temperature.

5. Two equal masses of air, both at 1000 mb, mix thoroughly. Their initial temperatures and mixing ratios are: $T_1 = 23.8\,°C$, $r_1 = 16.3 \text{ g kg}^{-1}$, $T_2 = -6.4\,°C$, $r_2 = 1.3 \text{ g kg}^{-1}$. Describe the final result, expressing: the temperature, the mixing ratio, if air is saturated or not and, if so, the liquid water content in g m^{-3}. Make all calculations analytically; if necessary, use a table of saturated vapor pressures or of saturation mixing ratios.

6. Two equal masses of air, one at $0\,°C$ and the other at $25\,°C$, both saturated but without any liquid water, mix thoroughly. The pressure is 1000 mb. Calculate the liquid water content of the resulting fog, in g m^{-3}. Use tables as needed, but no diagrams.

7. On a given occasion, convective activity becomes visualized by isolated cumuli whose bases are at 2000 m above the ground. Near the ground, the temperature and dew point have been constant for the last hour, with values $30\,°C$ and $13.3\,°C$, respectively. Would you say that the cumuli might have been formed by air rising from near the ground? Explain.

8. Complete the tephigram of Problem VI-1 by drawing three saturated adiabats: $\theta_{aw} = 250, 270$ and 290 K.

9. A parcel of cloud air rises, expanding adiabatically, from 1000 mb, $20\,°C$, until it reaches a temperature of $-40\,°C$. What is the final pressure? Make the following approximations:
   (a) Neglect the remaining mixing ratio of water vapor at $-40\,°C$,
   (b) Neglect the heat capacity of water (both liquid and vapor),
   (c) Neglect the partial pressure of water vapor against the total pressure, and
   (d) Assume a constant value of the latent heat of vaporization (use an average value between $-40\,°C$ and $+20\,°C$).

   Use tables as needed. Do not use a tephigram. The calculation must be made analytically.

10. Consider a mass of air which is aloft above the tropics, saturated with water vapor with respect to ice, at 230 mb and $-40\,°C$. Assume that it rises adiabatically until reaching the tropopause at 120 mb. Water vapor condenses to ice. What is the temperature (°C), within $1\,°$ of approximation, at the tropopause?

    The heat capacity of ice can be neglected; the water vapor pressure is negligible as compared with that of dry air; the water vapor at the tropopause can be neglected altogether (i.e., set $\cong 0$). You can use the Table of Constants in the book.

11. Saturated air rises adiabatically, water vapor condensing into water droplets while the air remains saturated. Make a simple sketch on a vapor pressure diagram (e, T) where you show the initial and final points, the path representing the process, and a second path to the same final point, consisting in
    (a) adiabatic expansion without condensation, followed by
    (b) condensation at constant pressure.

12. (a) Derive a formula for the (specific) work of expansion associated with a polytropic ascent of air, as a function of its temperature variation ΔT and the polytropic exponent n.

(b) Calculate that work, in J kg^{-1}, for $\Delta T = -10$ K and $n = 1.2$. How much is the corresponding absorbed heat, per kg?

13. Derive the expression for the final temperature distribution $T = f(p)$ acquired by an isothermal layer of temperature T_0 contained between the isobars p_1 and p_2, when it is thoroughly mixed vertically. Assume that there is no condensation.

14. With the data of Problem VI-4, find on a tephigram the adiabatic wet-bulb temperature T_{aw}, the potential adiabatic wet-bulb temperature θ_{aw} and the saturation temperature T_s and pressure p_s (temperature and pressure at which condensation starts, when expanded adiabatically). Give the temperatures in °C.

15. (a) Derive an expression for θ_{ae} as a function of T and p (r_w, e_w or T_s can be left as implicit functions of T, p). Use the usual approximate formula for saturated adiabats.

    (b) What is the value of θ_{ae} for saturated air at 800 mb, 0 °C, as obtained with the derived expression?

16. With the data of Problem IV-7, determine graphically, on a tephigram, r_w, θ, θ_v, p_s, T_s, T_{aw}, θ_{aw} and T_d. Compute T_{iw} and T_{ie}.

17. A Föhn blowing on the ground at 1000 mb has a temperature of 38 °C and a mixing ratio of 4 g kg^{-1}. Could this air be the same as that at the 1000 mb level on the windward side of the mountains, with a temperature of 21.5° and a mixing ratio of 10 g kg^{-1}? Could it be the same as that at the 800 mb level with 5 °C and 5 g kg^{-1}? Use the tephigram, and give the reasons for your answer.

18. A mass of air undergoes vertical displacements, during which precipitation falls out. If it can be assumed that no appreciable heat has been exchanged with the environment, what invariant parameter could be used to identify the mass through its transformations? What invariant parameter could be useful to identify a non-precipitating air mass moving horizontally over land and changing its temperature?

CHAPTER VIII

ATMOSPHERIC STATICS

In Chapter VII we have studied the behavior of individual air parcels when they undergo certain physical transformations. We now turn our attention to several general aspects related to the vertical stratification of the atmosphere and then, in the next chapter, to the vertical stability of the atmosphere. By 'vertical' we mean that the atmosphere will be considered above a certain location on the Earth's surface, generally taking into account neither the horizontal motions due to the Earth's rotation and to horizontal gradients of pressure, nor the large-scale vertical motions.

In this chapter we shall study hydrostatic equilibrium, several cases of ideal atmospheres, and the calculation of heights. We shall use only one coordinate z, in the vertical direction, increasing upwards, with its origin at mean sea level. We shall assume that the state variables remain constant for a constant z; that is, the isobaric, isothermal and equal humidity surfaces will be horizontal.

8.1. The Geopotential Field

Every system in the atmosphere is subject to the force of gravity. This is the resultant of two forces: (1) the gravitational attraction per unit mass $\mathbf{f}$, in accordance with Newton's universal law of gravitation, and (2) the centrifugal force, which results from choosing our frame of reference fixed to the rotating Earth; this much smaller component is equal (per unit mass) to $\omega^2 \mathbf{r}$, where ω is the angular velocity and r the distance from the axis of rotation. The vector sum $\mathbf{g}$ is the force of gravity per unit mass, or simply *gravity* (see Figure VIII-1):

$$\mathbf{g} = \mathbf{f} + \omega^2 \mathbf{r}. \tag{1}$$

If we call R the radius of the Earth (i.e., the distance from the center of mass to the surface), we must take into account that R varies with the latitude φ, due to the ellipticity of the Earth. The values are maximum and minimum at the Equator and at the Poles, respectively:

$R_{\text{Equator}} = 6378.1$ km

$R_{\text{Pole}} = 6356.9$ km.

From the inverse-square law of universal gravitation, it follows that the variation of $f = |\mathbf{f}|$ with altitude over a given location on the Earth's surface can be expressed by

$$f_z = f_0 \frac{R^2}{(R+z)^2} \tag{2}$$

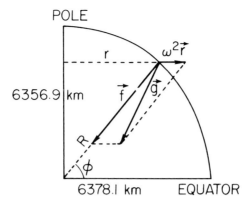

Fig. VIII-1. Forces in the geopotential field.

where f_0 is the value of f at mean sea level and f_z is the value at an altitude z above mean sea level. Thus

$$f_z = f_0\left(1 + \frac{z}{R}\right)^{-2} = f_0\left(1 - 2\frac{z}{R} + \ldots\right) \cong f_0\left(1 - 2\frac{z}{R}\right). \quad (3)$$

For the purpose of introducing R into Equation (3), an average value can be used and thus

$$f_z \cong f_0(1 - 3.14 \times 10^{-7} z) \quad (4)$$

with sufficient approximation (z in meters).

The gravity g at sea level varies with latitude due to the centrifugal force and to the ellipticity. As $\omega^2 r \ll f$, the angle between **g** and **f** is very small and we can write from the cosine theorem (see Figure VIII-1)

$$g_{\varphi,0} = f_0\left[1 + \left(\frac{\omega^2 r}{f_0}\right)^2 - 2\frac{\omega^2 r}{f_0}\cos\varphi\right]^{\frac{1}{2}} \cong f_0 - \omega^2 r \cos\varphi \quad (5)$$

where $g_{\varphi,0}$ means gravity at latitude φ and sea level. Or, considering that $r = R\cos\varphi$

$$g_{\varphi,0} = f_0 - \omega^2 R \cos^2\varphi. \quad (6)$$

At $\varphi = 45°$,

$$g_{45,0} = f_0 - \tfrac{1}{2}\omega^2 R. \quad (7)$$

Eliminating f_0 between Equations (6) and (7), we find

$$g_{\varphi,0} = g_{45,0} - \tfrac{1}{2}\omega^2 R \cos 2\varphi. \quad (8)$$

As R is approximately constant, Equation (8) indicates that the centrifugal term is proportional to $\cos 2\varphi$; it amounts to -1.7 cm s^{-2} at the equator, 0 at 45° and $+1.7$ cm s^{-2} at the poles.

The variation of g with latitude due to ellipticity includes the effect of variation of distance from the mass center (which attains a maximum of 21 km between the equator and the poles) and the effect of a non-spherical distribution of mass. The former would imply a difference in gravity of 6.6 cm s^{-2} between the poles and equator, and the latter partially compensates this difference. Both may be taken into account together with the centrifugal term by adjusting the coefficient of $\cos 2\varphi$ in Equation (8). Thus the following approximate formula for the dependence of g on latitude may be written:

$$g_{\varphi,0} = g_{45,0}(1 - 0.00259 \cos 2\varphi). \tag{9}$$

We consider now that because g differs little from f the same correction factor for altitude of formula (4) (expression between brackets) can be applied to it:

$$g_{\varphi,z} = g_{\varphi,0}(1 - 3.14 \times 10^{-7} z) \tag{10}$$

and introducing Equation (9), we finally obtain

$$g_{\varphi,z} = g_{45,0}(1 - a_1 \cos 2\varphi)(1 - a_2 z) \tag{11}$$

where

$$a_1 = 2.59 \times 10^{-3}$$

$$a_2 = 3.14 \times 10^{-7} \text{ m}^{-1}$$

$$g_{45,0} = 9.80616 \text{ m s}^{-2}.$$

As Equation (11) indicates, gravity at mean sea level varies from 9.78 m s^{-2} at the equator to 9.83 m s^{-2} at the poles.

More complicated expressions have been developed and must be used for accurate estimates of the gravity, which also depends slightly on local topography. Meteorologists use the so-called *meteorological gravity system*, which gives values very slightly different from the *Potsdam system*, widely used in geodesy. These differences need not concern us here. More accurate formulas than (11) to calculate local values of the acceleration of gravity can be consulted in the WMO Tables (see Bibliography).

A *standard value* at sea level g_0 has also been adopted for reference. It is

$$g_0 = 9.80665 \text{ m s}^{-2}. \tag{12}$$

If a unit mass moves in the gravitational field, the force of gravity performs a work

$$\delta w = \mathbf{g} \cdot \mathbf{dr} = g \cos \theta \, dr \tag{13}$$

where $\mathbf{dr}$ is the displacement, θ the angle it forms with $\mathbf{g}$, and the dot between the vectors indicates a scalar product.

Experience shows that when the mass returns to its original position

$$\oint \mathbf{g} \cdot \mathbf{dr} = 0. \tag{14}$$

This property is expressed by saying that the gravitational field is *conservative*. It follows from Equation (14) that δw is an exact differential that defines w as a point function. It is preferred, however, to define it with opposite sign, and it is called the *geopotential* ϕ. The conservatism of the gravitational field is also expressed by saying that **g** is a force derived from a potential.

We shall have, in general

$$\phi = \phi(x, y, z),$$

$$d\phi = -\mathbf{g} \cdot d\mathbf{r} \quad \text{(exact differential)}, \tag{15}$$

$$\Delta\phi = -\int \mathbf{g} \cdot d\mathbf{r},$$

where $\Delta\phi$ is independent of the integration path. As ϕ depends only on the altitude z,* these expressions simplify to

$$\phi = \phi(z)$$

$$d\phi = g\, dz \tag{16}$$

$$\Delta\phi = \int g\, dz \cong g\Delta z.$$

The second one says that the gravity is given by the geopotential gradient. The last approximate equation indicates that, as g varies little with z, it may be taken for many purposes as a constant.

As a point function, ϕ is defined except for an additive constant. We fix it by choosing zero geopotential at mean sea level:

$$\phi(0) = 0, \tag{17}$$

so that

$$\phi(z) = \int_0^z g\, dz \cong gz. \tag{18}$$

8.2. The Hydrostatic Equation

In a state of equilibrium, the force of gravity is everywhere balanced by the pressure forces, whose resultant is opposite to it. Let us consider a layer of thickness dz in a column of unit area (Figure VIII-2). A force p acts on its base, directed upwards,

---

* If z is measured along a line of gravity force, i.e., along a line parallel at every point to **g**, it will follow a slightly curved line. This deviation from a straight line is however negligible for most purposes.

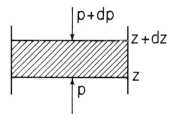

Fig. VIII-2. Pressure variation in vertical.

while a downward force $p+dp$ acts on the top. Therefore a net force $-dp$ is acting upwards on a mass $\varrho\, dz$. The force of gravity is $g\varrho\, dz$. As they must cancel:

$$dp = -g\varrho\, dz$$

or

$$\frac{dp}{dz} = -g\varrho. \tag{19}$$

This is the hydrostatic equation for the simple case when $p = p(z)$. The isobaric surfaces, as well as the equipotential surfaces, are considered here to be horizontal.

If we compare Equations (16) and (19), we have for hydrostatic equilibrium the equivalent expressions:

$$dp = -\varrho\, d\phi. \tag{20}$$

and

$$d\phi = -v\, dp. \tag{21}$$

Integration of Equation (21) gives:

$$\Delta\phi = -\int_1^2 v\, dp = -\int_1^2 RT\, d\ln p = -R_d \int_1^2 T_v\, d\ln p. \tag{22}$$

In order to perform this integration, we must know the virtual temperature T_v as a function of the pressure p. We shall consider this problem in Section 5 for a particular kind of atmosphere, and in Section 11 for the general case.

8.3. Equipotential and Isobaric Surfaces. Dynamic and Geopotential Height

The geopotential field may be described by the equipotential surfaces; the thickness, h_ϕ, of a layer of unit geopotential difference, $\Delta\phi$, will be given by:

$$\Delta\phi = gh_\phi = 1 \text{ unit of geopotential.} \tag{23}$$

ϕ has dimensions of energy per unit mass or specific energy. In the MKS system the units are J kg^{-1} = m^2 s^{-2}. Thus, with $g \cong 9.8$ m s^{-2}, Equation (23) gives

$$h_\phi \cong 0.102 \text{ m}. \tag{24}$$

Therefore approximately 1.02 m (the exact value depending on that of g) correspond to a difference in geopotential of 10 J kg^{-1}; this equivalence suggested at one time the use of another unit for the specific energy, called the *dynamic metre*, which may be abbreviated dyn-m and was defined by

$$1 \text{ dyn-m} = 10 \text{ J kg}^{-1}. \tag{25}$$

The dynamic kilometre (dyn-km) or other similar units could also be used; they bear the same relations among themselves as the corresponding units of length. The value of ϕ in these units has in the past been frequently referred to as the *dynamic height* or *dynamic altitude*. They are seldom encountered in meteorology at the present time.

It is now customary in meteorology to express the geopotential as *geopotential altitude*. As a physical quantity this is again identically the geopotential, but expressed in units called *standard geopotential metres* or, more simply, *geopotential metres* (gpm). The conversion factor has the numerical value of the standard gravity $|g_0| = 9.80665$:*

$$1 \text{ gpm} = |g_0| \text{ J kg}^{-1} \tag{26}$$

Thus the geopotential can be written

$$\phi = \int_0^z g \, dz \, (\text{J kg}^{-1}) = \frac{1}{|g_0|} \int_0^z g \, dz \, (\text{gpm}) = \frac{\bar{g}}{|g_0|} z \, (\text{gpm}) \tag{27}$$

and is called 'geopotential altitude' if expressed in gpm or any multiple of this unit (such as gpkm = geopotential kilometre). In the last expression, $\bar{g}$ is the average value of g in the integration interval. As $\bar{g}/|g_0|$ is numerically very close to unity (within a fraction of 1% anywhere in the troposphere), it is obvious from Equation (27) that the numerical values of the geopotential in gpm (i.e., the geopotential altitude) and of the altitude in m are almost equal.

Summarizing the equivalent expressions of geopotential in different units, we have:

$$\phi = \bar{g}z \text{ J kg}^{-1} = \frac{\bar{g}}{|g_0|} z \text{ gpm} = \frac{\bar{g}}{10} z \text{ dyn-}m \tag{28}$$

where $\bar{g}z = \int_0^z g \, dz$ and z is given in metres.

It must be stressed that neither the dynamic meter nor the geopotential metre are units

---

* Until 1971, the convention used in meteorology was to take a conversion factor equal to 9.8 (J kg^{-1}) gpm^{-1} (exactly).

of length, but units of geopotential, i.e., of specific energy. Neither is the 'dynamic altitude' or the 'geopotential altitude' a length, but identically the geopotential, the word 'altitude' merely referring to the type of specific energy units used (dyn-m or gpm, respectively).

If we introduce the expression (10) to calculate the average value g between mean sea level and z, we obtain

$$\bar{g} = \frac{1}{z}\int_0^z g\,dz = \frac{g_{\varphi,0}}{z}\int_0^z (1 - 3.14 \times 10^{-7} z)\,dz$$

$$= g_{\varphi,0}(1 - 1.57 \times 10^{-7} z) \qquad (29)$$

and

$$\phi = \frac{g_{\varphi,0}}{|g_0|} z(1 - 1.57 \times 10^{-7} z) \qquad (30)$$

(z in metres).

In general, the ratio $\bar{g}/|g_0|$ in formula (27) will differ from unity by less than 0.2%. It is customary in meteorology to use the geopotential height rather than geometrical height for the representation of the state of the free atmosphere, for a number of reasons, theoretical as well as practical. In the first place, in most dynamic equations the differential of height is associated with the acceleration due to gravity as a product, and $g\,dz$ can always be replaced by $|g_0|\,d\phi$ with ϕ in geopotential altitude units. This is particularly true if isobaric coordinates (x, y, p, t) are used (as is now standard), where position in the vertical is defined by the pressure. In the second place, true or geometrical height is never required and is seldom measured; this is partly because of the central position of pressure (or related parameters) as an indicator of position in the vertical. Finally, the computational advantages of ϕ over z virtually necessitate the use of geopotential height.

When the foot is used as a unit of length, the exact equivalences 1 ft = 0.3048 m and 1 gpft = 0.3048 gpm are adopted, where gpft stands for *geopotential foot*.

The thickness h_p, between isobaric surfaces separated by a unit pressure difference, can be similarly calculated from Equation (19):

$$\Delta p = g\varrho h_p = 1 \text{ unit of pressure}$$

$$h_p = \frac{1}{g\varrho} = \frac{v}{g}. \qquad (31)$$

But in this case v, and therefore h_p, vary rapidly with p, and therefore with height. At 1000 mb, v is about 0.8 m³ kg⁻¹, and $h_p = 0.08$ m (for 1 Pa = 0.01 mb); it follows that in the layers near to the ground, the pressure drops with height at a rate of 100 mb for 800 m. As we rise in the atmosphere, the unit layers become thicker.

By comparing Equations (23) and (31), we obtain

$$h_p = vh_\phi. \qquad (32)$$

8.4. Thermal Gradients

The vertical temperature gradient or lapse rate may be defined by the derivative

$$\beta = -dT/dz. \tag{33}$$

However, it is also convenient to define it in the atmosphere with respect to the geopotential:

$$\gamma = -\frac{dT}{d\phi} = -\frac{1}{g}\frac{dT}{dz} = \frac{\beta}{g}. \tag{34}$$

We shall use this last definition.

It should be noticed that altough both parameters are proportional and have close numerical values (if dynamic or geopotential heights are used for ϕ), they are two physical quantities of different kind. The first definition (β) gives the decrease in the absolute temperature with height, and its units are K m^{-1} (in MKS system). γ gives the decrease of temperature with the geopotential, and is measured in K s^2 m^{-2} in MKS system, although it is customarily expressed, for convenience, in degrees per unit of geopotential height (K gpm^{-1} or K gpft^{-1}).

Starting from the virtual temperature, a lapse rate of virtual temperature γ_v is similarly defined:

$$\gamma_v = -\frac{dT_v}{d\phi}. \tag{35}$$

Other expressions for γ_v may be derived if we take into account Equations (18), (21) and the gas law:

$$\gamma_v = -\frac{1}{g}\frac{dT_v}{dz} = \frac{1}{v}\frac{dT_v}{dp} = \frac{1}{R_d}\frac{d\ln T_v}{d\ln p}. \tag{36}$$

We remark now that these derivatives and differentials may refer: (1) to the variations undergone by an air mass during a process, for instance during an adiabatic ascent, and (2) to the variations in the values of the static variables along the vertical for an atmosphere at rest. We shall call the first ones *process derivatives* or *process differentials*, while we shall refer to the latter as *geometric derivatives* or *geometric differentials*. We shall only be concerned in this chapter with geometric variations.

8.5. Constant-Lapse-Rate Atmospheres

We shall now consider the case of an atmosphere with constant lapse rate γ_v (of virtual temperature), for which T_v decreases proportionally with ϕ. This is a particularly important case inasmuch as it is the obviously simplest way to approximate a real atmosphere. As in general, in the troposphere, temperature decreases with height, γ_v is usually a positive quantity.

We write for $z=0$ (mean sea level):

$$T_v = T_{v_0}, \quad p = p_0, \quad \phi = 0.$$

We find $T_v = f(\phi)$ by integrating Equation (35) from mean sea level to any height:

$$T_v = T_{v_0} - \gamma_v \phi$$

or

$$\frac{T_v}{T_{v_0}} = 1 - \frac{\gamma_v \phi}{T_{v_0}} \tag{37}$$

i.e., the virtual temperature decreases linearly with the geopotential, or, if we neglect the small variation of g, with the altitude.

From Equation (36) we obtain:

$$d \ln T_v = R_d \gamma_v \, d \ln p, \tag{38}$$

which contains the hydrostatic equation, and integrated gives $T_v = f(p)$:

$$\frac{T_v}{T_{v_0}} = \left(\frac{p}{p_0}\right)^{R_d \gamma_v} \tag{39}$$

and by eliminating (T_v/T_{v_0}) between Equations (37) and (38), we obtain $p = f(\phi)$:

$$p = p_0 \left[1 - \frac{\gamma_v \phi}{T_{v_0}}\right]^{1/R_d \gamma_v}. \tag{40}$$

Thus, Equations (37), (39) and (40) relate the variables T_v, p and ϕ. Equations (39) and (40) are the result of integrating the hydrostatic equation for the particular case $\gamma_v = $ const. It may be noticed that, dimensionally, $[\gamma_v] = 1/[R_d]$.

If we consider dry air (for which $T_v = T$) and compare formula (39) with Chapter II, formula (61), we see that the formulas are equivalent if we set $k = (n-1)/n = R_d \gamma_v$, i.e., the formulas for an atmosphere with constant lapse rate are similar to those for polytropic processes. However, they describe here the geometric distribution of temperature and pressure rather than variations during a process.

We can see from Equations (37) and (40) that p and T_v become 0 for $\phi_1 = T_{v_0}/\gamma_v$. This is therefore called the *limiting geopotential height of an atmosphere with constant lapse rate* (we prefer this terminology to the simpler term geopotential height, which is open to some ambiguity). For this idealized model atmosphere, there will not be any air for $\phi > \phi_1$.

We shall now consider three special cases of constant-lapse-rate atmospheres.

8.6. Atmosphere of Homogeneous Density

If $\gamma_v = 1/R_d = 34.2$ K gpkm^{-1}, from Equation (38) and the gas law relation d ln $T_v =$

$= d \ln p + d \ln v$, it follows that $dv = 0$. This atmosphere has therefore a homogeneous density. It is sometimes called the 'homogeneous atmosphere', although only density is constant, while temperature and pressure vary with height.

Equations (37), (39) and (40) simplify to

$$T_v = T_{v_0} - \phi/R_d \tag{41}$$

$$T_v/T_{v_0} = p/p_0 \tag{42}$$

$$p = p_0(1 - \phi/R_d T_{v_0}) \tag{43}$$

where the second one is simply the gas law for constant density; and the limiting geopotential height is

$$\phi_1 = R_d T_{v_0}.$$

If $T_{v_0} = 273$ K, $\phi_1 = 7990$ gpm. We shall see later that the lapse rate never reaches this value in the real atmosphere, except for very shallow layers over a strongly heated surface.

8.7. Dry-Adiabatic Atmosphere

We consider now an atmosphere with a lapse rate:

$$\gamma_v = 1/c_{p_d} = \gamma_d = 9.76 \text{ K gpkm}^{-1}.$$

Equation (39) becomes

$$T_v = T_{v_0}\left(\frac{p}{p_0}\right)^{\varkappa_d} \tag{44}$$

which coincides with one of Poisson's equations describing the adiabatic expansion of dry air. The geometric distribution of temperature for this static atmosphere is therefore described by the curve followed by the adiabatic expansion process of a parcel of dry air. It will be characterized by the virtual potential temperature:

$$\theta_v = T_{v_0}\left(\frac{1000 \text{ mb}}{p_0}\right)^{\varkappa_d} \tag{45}$$

which is constant throughout the whole atmosphere, as can be verified by introducing Equation (44) into the expression of θ_v for any T_v, p.

Equations (37) and (40) become:

$$T_v = T_{v_0} - \phi/c_{p_d}, \tag{46}$$

and

$$p = p_0\left(1 - \frac{\phi}{c_{p_d} T_{v_0}}\right)^{1/\varkappa_d} \tag{47}$$

and the limiting geopotential height is:

$$\phi_1 = c_{pd} T_{v0},$$

If $T_{v0} = 273$ K, $\phi_1 = 27\,950$ gpm.

8.8. Isothermal Atmosphere

We assume:

$$\gamma_v = 0, \qquad T_v = T_{v0} = \text{const.}$$

Integration of Equation (21) gives:

$$\phi = -\int v\,dp = -R_d T_{v0} \ln \frac{p}{p_0} \tag{48}$$

or

$$p = p_0 e^{-\phi/R_d T_v} = p_0 e^{-\phi/RT} \tag{49}$$

Therefore, in this case, pressure decreases exponentially with height, and tends to 0 when $z \to \infty$. The limiting geopotential height of this atmosphere is thus infinite.

The expression (49) can also be written

$$p = p_0 e^{-gM_d z/R^* T_v} = p_0 e^{-gMz/R^* T} = p_0 e^{-z/H} \tag{50}$$

where M_d is the average molecular weight of dry air, M is the average molecular weight of the air in the atmosphere and we have defined the parameter

$$H = \frac{R^* T_v}{g M_d} = \frac{R^* T}{g M} \tag{51}$$

which is called the *scale height* and represents the altitude at which the ground pressure becomes reduced by the factor e^{-1}. H is a constant for an isothermal atmosphere, if the variation of g is neglected. When T varies with height, H can no longer be considered constant. In that case one can speak of a 'local' scale height; for instance, the geopotential height of the atmosphere of homogeneous density (Section 6) corresponds to the value at the ground of the scale height: $H_1 = \phi_1/g = R^* T_{v0}/gM$.

8.9. Standard Atmosphere

Of the particular cases which we have considered, the atmosphere of homogeneous density has only a theoretical interest, and the isothermal atmosphere is only applicable to layers with a constant temperature. In the lower stratosphere, this is frequently a useful approximation, especially at high latitudes, but in the troposphere is seldom valid except for relatively thin layers. In this latter domain, the adiabatic atmosphere is of more practical importance, since it gives, as we shall see, an upper limit for the value of the lapse rate of a vertically stable atmosphere, and has also the temperature distribution of a vertically-mixed layer (cf. Chapter VII, Section 12). But the real atmosphere has always on the average lower values of γ and, although this is in general not constant with height, we may define an atmosphere with a constant lapse rate which approximates the real average case (for the troposphere).

Thus several 'standard atmospheres' have been defined, which are particularly important in aeronautics, where they are used for reference as an approximation to the real atmosphere, and for calibrating and using altimeters.

We give here the basic definitions of the standard atmosphere adopted by the International Civil Aviation Organization (ICAO). These are stated using an altitude H' corrected for the variation of g, according to the expression

$$H = \frac{1}{g_0} \phi = \frac{1}{g_0} \int_0^z g \, dz. \tag{52}$$

Here g_0 is the standard value of gravity, while g is the actual value, depending on latitude and altitude. The formula is similar to Equation (27), but here g_0 retains its dimensions as an acceleration, and therefore H is a real length. Obviously H is numerically equal to the geopotential ϕ expressed in geopotential height units. As we have seen in Section 1, g decreases by about 0.3% for a 10 km increase of altitude. If we disregard the variations of g, $H \cong z$. The conditions defining the standard atmosphere are:

(1) Atmosphere of pure dry air with constant chemical composition in the vertical, with mean molecular weight $M = 28.9644$ (C^{12} scale).

(2) Ideal gas behavior.

(3) Standard sea-level value of the acceleration due to gravity: $g_0 = 9.80665$ m s^{-2}.

(4) Hydrostatic equilibrium.

(5) At mean sea level, the temperature is $T_0 = 15\,°C = 288.15$ K and the pressure $p_0 = 1013.25$ mb $= 1$ atm.

TABLE VIII-1
Standard atmosphere

p (mb)	H (m)	T (°C)
1013.25	0	15.0
1000	110	14.3
900	990	8.6
800	1 950	2.3
700	3 010	− 4.6
600	4 200	− 12.3
500	5 570	− 21.2
400	7 180	− 31.7
300	9 160	− 44.5
226.3	11 000	− 56.5

(6) For values of H up to 11 000 m above mean sea level (tropopause) the lapse rate is constant and given by $\beta_0 \doteq -dT/dH = 6.5$ K km^{-1}.

(7) For altitudes $H \geqslant 11\,000$ m (stratosphere) and up to 20 000 m, the temperature is constant and equal to $-56.5\,^\circ$C. Then the lapse rate becomes -1.0 K km^{-1}, up to 32 000 m.

Table VIII-1 gives some values of p, H and T for this standard atmosphere.

According to Equations (37), (39) and (40), the relations between T, p and ϕ (or H), are given for the standard atmosphere up to 11 000 m by

$$T = T_0 - \beta_0 H = 288.15 - 6.5 \times 10^3 H \tag{53}$$

$$T = T_0 \left(\frac{p}{p_0}\right)^{R_d \beta_0/g_0} = 288.15 \left(\frac{p}{1013.25}\right)^{0.1903} \tag{54}$$

$$p = p_0 \left(1 - \frac{\beta_0 H}{T_0}\right)^{g_0/R_d \beta_0}$$

$$= 1013.25(1 - 2.255 \times 10^{-5} H)^{5.256} \tag{55}$$

where T is given in K, p in mb and H in m. It may be noticed that, as H and the geopotential altitude are numerically equal, and so are γ_0 and β_0, if the former is the corresponding lapse rate of the standard atmosphere expressed in K gpm^{-1}, ϕ in gpm can be written for H in the previous formulas, without further change ($\beta_0 H = \gamma_0 \phi$, where $\gamma_0 = 6.5$ K gpkm^{-1}).

8.10. Altimeter

Because of its practical importance, we shall describe the altimeter used in airplanes to determine the height at which they are flying, and the corrections that must be made to its readings.

The altimeter is an aneroid barometer with two scales, which we shall call the *main scale* and the *auxiliary scale*. They correspond basically to two different ways of measuring the pressure. A mechanical transmission measures the effect of pressure on the barometer by the position of hands moving over the main scale. This is schematically represented in Figures VIII-3a–c by the position of a pointer. An alternative way of measuring the pressure is by using the auxiliary scale and a compensation method. The auxiliary scale is actually covered except for a window with an index showing the reading matched to the zero of the main scale (see figure). It is graduated in pressure units, in such a way that the reading is the correct pressure on the barometer when the hands read zero on the main scale. This situation is obtained by shifting the scales appropriately, and is indicated in Figure VIII-3a; in the real instrument the main scale and the window are actually fixed and it is the auxiliary scale and the hands that rotate simultaneously (rather than the main scale, as shown for convenience in the figure). The real scales of the instrument look like the sketch in Figure VIII-3d.

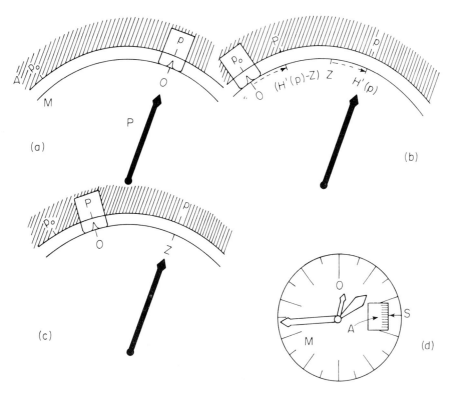

Fig. VIII-3. Altimeter scales.

The reading on the auxiliary scale matched to the zero on the main scale is called the *altimeter setting* or *Kollsman number*; this is p, p_0 and P in Figure VIII-3a, b and c, respectively.

The main scale is not graduated in pressure units, but in altitude units, related to the pressure by Equation (55). Therefore, if the altimeter setting is $p_0 = 1013.25$ mb, the hands will show the correct altitude on the main scale when used in a standard atmosphere. In particular, they will show zero altitude at 1013.25 mb. This is the setting in Figure VIII-3b. The reading $H(p)$, corresponding to any pressure p, is called the *pressure altitude* or *standard altitude*. It is the correct altitude of the isobar p in the standard atmosphere and coincides therefore with the value of H from Equation (55). For any particular setting different from p_0, the hands on the main scale at any moment indicate the difference of actual pressure from that of the setting, in altitude units H; this reading will be the real difference in altitude if the atmosphere is standard.

If the altimeter is set at p_0 while being used in an atmosphere different from the standard, in that zero altitude corresponds to another pressure P, the readings will be in error. Thus if the real altitude of a station with pressure p is Z, clearly $Z \neq H(p)$ unless the atmosphere is standard. In order to match the readings Z and p, we have to displace the zero altitude (Figure VIII-3c) so that it will correspond to a pressure P in the standard atmosphere for an altitude $(H(p) - Z)$. If an aircraft obtains information about the value of P from an airport and sets its altimeter accordingly, the reading will give the altitude of the airport when landing.

The procedure for obtaining the altimeter setting P is therefore:

(1) Obtain $H(p)$ from the actual pressure p at the station, by using a standard atmosphere table or Equation (55).

(2) Subtract the true altitude of the field Z.

(3) Obtain the altimeter setting P from $(H(p) - Z)$ through a standard atmosphere table or Equation (55).

This pressure correction is equivalent to approximating the atmosphere over the station by a standard atmosphere where a constant has been added to all pressures, the constant being chosen so as to make p correspond to Z.

Conversely, if instead of setting the altimeter to P, it is set with zero altitude at p_0, the correction to be added to any reading z above the station would be $D = z - H(p)$, sometimes called the 'D factor'.

The pressure correction has to be determined carefully as the values of z are very sensitive to small pressure variations. According to Equation (55), an error of 1 mb in p will give an error of 8 to 9 m in z.

To explain the temperature correction, let us assume that the station is at sea level, and its temperature is T_0' instead of $T_0 = 15\,°C$. The average lapse rate of the atmosphere β up to the flying level will also be different in general from the standard value β_0. If the altimeter is set at the standard pressure p_0, its reading z' will be given by Equation (55), which may be written

$$p = p_0\left(1 - \frac{\beta_0 z'}{T_0}\right)^{g_0/R_d\beta_0}. \tag{56}$$

Here p is the actual pressure and the difference between actual (z) and corrected (H) altitudes is neglected, for simplicity. If we approximate the real atmosphere by an atmosphere with constant lapse rate β, the relation between the pressure p and the real altitude z will be given (from Equation (40)) by

$$p = p_0\left(1 - \frac{\beta z}{T_0'}\right)^{g_0/R_d\beta}.$$

Equating these two expressions and solving for z, we obtain

$$z = \frac{T_0'}{\beta}\left[1 - \left(1 - \frac{\beta_0 z'}{T_0}\right)^{\beta/\beta_0}\right]. \tag{57}$$

If z is not large, $\beta_0 z'/T_0$ and $\beta z/T_0'$ may be considered small relative to unity,* the exponential within the bracket in Equation (57) can be developed and only the first order term retained:

---

* Even for large z this ratio will not exceed about 0.25.

$$z = \frac{T_0'}{\beta}\left[1 - 1 + \frac{\beta}{\beta_0}\frac{\beta_0 z'}{T_0} - \cdots\right] \cong \frac{T_0' z'}{T_0} \tag{58}$$

or

$$z = z'\left(1 + \frac{\Delta T}{T_0}\right) \tag{59}$$

which gives the correction by a simple formula ($\Delta T = T_0' - T_0$).

It may be remarked that differences in lapse rate from the standard value have little influence on the correction, because they only enter in second-order terms in the previous development, so that they do not appear in the first-order approximation (59).

This correction may be in appreciable error if there are discontinuities in the atmosphere, as when the aircraft is flying above a frontal or a subsidence inversion. It should also be remarked that the temperature correction is much less critical than the pressure one, because it is proportional to the altitude and thus it becomes zero on landing, whereas an error in the setting will give an additive difference at all altitudes. With respect to the error during flight, for a given difference of pressure between ground and the aircraft, the altitude over the ground will be larger than the indication if $T_0' > T_0$, because the intermediate atmosphere is less dense than the standard.

8.11. Integration of the Hydrostatic Equation

Formulas (39) and (40) give the result of integrating the hydrostatic equation for the particular case in which γ_v is a constant. The important problem arises now of finding a convenient method for obtaining the pressure (or the temperature) as a function of the geopotential ϕ (or of the altitude z) in any real atmosphere; that is, a convenient method for integrating the hydrostatic equation. This is easily done with an aerological diagram, as we shall see, by a summation over successive atmospheric layers.

Let us consider formula (22):

$$\Delta\phi = -\int_1^2 v\,dp = -\int_1^2 RT\,d\ln p = -R_d\int_1^2 T_v\,d\ln p. \tag{60}$$

This is the same integral that appears in Chapter VI, Equation (17), whose graphical determination was explained in Chapter VI, Section 11. It should be realized that its meaning is now different. Here $\Delta\phi$ depends on the geometric structure, so that dp is a change of pressure with height in a static atmosphere; in Chapter VI we were considering a reversible process undergone by an air parcel. However, the value of the integral will be the same in both cases, provided the path of the parcel is represented by the same curve describing the geometric change of properties in the static atmosphere. Therefore, we can apply the same methods described before. Thus

$$-\int_1^2 T_v\,d\ln p = \sum_{em} \tag{61}$$

according to Chapter VI, formula (20) and with the same meaning of the area $\sum_{em}$, which can be determined by any of those methods. The mean isotherm method will give

$$\Delta\phi = - R_d \overline{T}_v \ln \frac{p_2}{p_1} = R_d \sum_{em} \tag{62}$$

(where we might substitute $\overline{RT}$ for $R_d \overline{T}_v$). For each pair of chosen isobars p_1, p_2 this depends only on $\overline{T}_v$ and can be printed for convenience on the diagram. The mean adiabat method gives the formula (cf. Equations (60), (61) and Chapter VI, Equation (26)):

$$\Delta\phi = c_p(T_1''' - T_2''') \cong c_{pd}(T_1''' - T_2''')$$
$$= 1005(T_1''' - T_2''') \text{ J kg}^{-1} = 102.5(T_1''' - T_2''') \text{ gpm} \tag{63}$$

where T_1''' and T_2''', as in Chapter VI, Section 11, are the temperatures of the intersections of the mean adiabat with the isobars p_1 and p_2.

If a tephigram is used, we could apply again the relation

$$\Delta\phi = -\int_1^2 v\,dp = -\Delta h + q = -c_p\Delta T + c_p \int_1^2 T\,d\ln\theta =$$
$$= c_p(-\Delta T + \sum_{te}) \tag{64}$$

(cf. Chapter VI, Equations (18) and (28)). But it will be more convenient to use the same methods as for the emagram, as explained in Chapter VI, Section 11 (cf. Chapter VI, Equation (31)).

With these procedures, the curve $\phi = f(p)$ or $z = f(p)$ can be computed and plotted on a diagram, on the basis of the state curve $T = f(p)$ representing the vertical structure of the atmosphere (the sounding). The procedure will follow the construction of a table such as Table VIII-2. The starting value will be the known altitude z_s of the station, where the pressure is p_s. The sounding is then divided in layers between succes-

TABLE VIII-2
Computation of altitude

i	p_i	$\Delta_i z$	$z_i = z_s + \sum_1^i \Delta_i z$
0	p_s		z_s
1	p_1	$\Delta_1 z$	z_1
2	p_2	$\Delta_2 z$	z_2
3	p_3	$\Delta_3 z$	z_3
—	—	—	—

sive isobars p_i, generally at intervals of 50 or 100 mb. Equation (60), applied to the first layer between p_s and p_1, will give $\Delta_1\phi$ and therefore $\Delta_1 z = z_1 - z_s$; this computation is performed graphically by any of the procedures described. The altitude of the isobar p_1 will be given by $z_1 = z_s + \Delta_1 z$. Then the second graphical integration is performed on the second layer; the result, $\Delta_2 z$, added to z_1, will give the altitude z_2 of the isobar p_2. And the procedure is continued along the whole sounding. The altitudes z_i can then be plotted on the same diagram, as a function of p_i, using an auxiliary scale for z.

The altitude curve has only a small curvature on an emagram or a tephigram, because the temperature varies slowly and thus ϕ or z is roughly proportional to $\ln p$. The vertical dimension, normal to isobars on the diagram, is therefore roughly proportional to height in the atmosphere.

It must be emphasized that these calculations are to be performed with the curve of virtual temperature as a function of pressure. In aerological practice, the relative humidity at any level is converted to the corresponding mixing ratio, and from the latter is computed the so-called virtual temperature increment, $\Delta T_v = T_v - T$, which is given from Equations (72) and (75) of Chapter IV, with sufficient accuracy, as $\Delta T_v \cong 0.6 r T$.

It may be of interest to compute what would be the pressure at mean sea level at the location of a station of height ϕ_s, should the atmosphere extend to $\phi = 0$; or to estimate the pressure at any other close level. This problem arises in routine meteorological practice in the production of mean-sea-level (m.s.l.) pressure values for a 'surface' chart. Since pressure values (at a constant level) are useful in the diagnosis of wind fields (at that level) and in the synoptic appreciation of weather systems, it is necessary to 'reduce' ground pressures to a reference level (chosen as mean sea level) in order to construct meaningful isobars. The procedure to be described below is reasonably satisfactory if station heights and temperatures are themselves relatively smooth fields in the horizontal, especially at elevations not in excess of 1 km. Let us consider the above problem, assuming that $\phi_s > 0$. The computation is made by considering a fictitious atmosphere from $\phi = 0$ to ϕ_s; to it is assigned a lapse rate equal to $\gamma_d/2$, as a reasonable mean value. The temperature T_0 and pressure p_0 at mean sea level are then computed from

$$T_0 = T_s + \frac{\gamma_d}{2}\phi_s \tag{65}$$

$$p_0 = p_s\left(1 + \frac{\gamma_d\phi_s}{2T_s}\right)^{2/\varkappa_d}. \tag{66}$$

Here it is customary to take for T_s an average value in such a way as to avoid the strong variations close to the ground, due to its daily warming and cooling by radiation. These variations are shown in Figure VIII-4. T_s is taken as the average between the actual value and that of 12 hours before:

$$T_s = \frac{T_{s,\text{actual}} + T_{s,-12\text{hs}}}{2}. \tag{67}$$

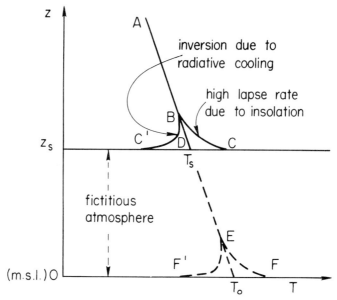

Fig. VIII-4. Computation of 'mean-sea-level pressures' by considering a fictitious atmosphere.

The additional temperature for reduction to m.s.l. is given by the dotted line of lapse rate $\gamma_d/2$. To compensate for the substitution of BC or BC' by BD, which alters the contribution of this layer to the pressure, a profile ABDEF or ABDEF' is assumed, whereby the peculiar stratification near the ground is transferred to the fictitious ground at sea level.

PROBLEMS

1. Find the dry adiabatic lapse rate, in K gpkm^{-1}, for an atmosphere composed entirely of Ar. The atomic weight of Ar is 39.9.
2. An aircraft is flying at 1000 m of altitude. The temperature T_v is 15 °C and the pressure p, 894 mb. (a) Assuming that the mean lapse rate γ_v below the aircraft is 8.5 K gpkm^{-1}, find the pressure p_0 at mean sea level. (b) Derive an expression to estimate the relative error in p_0 for a given error in γ_v. (c) What is the error in p_0 if the real mean lapse rate if 6.5 K gpkm^{-1}?
3. The sounding of an atmosphere between 850 and 700 mb is represented by the points:

p(mb)	T(°C)
850	8.0
800	3.0
750	4.0
700	1.0

Plot these data on a tephigram, and compute the thickness of the layer by the methods of the mean isotherm and the mean adiabat.

4. The layer between 1000 and 900 mb has a constant lapse rate, with $+3\,°C$ at the base and $-3\,°C$ at the top. Derive its geopotential thickness by several approximation methods, using a tephigram, and express it in gpm, in dyn-m and in $J\,kg^{-1}$.

5. What is the height where the horizontal pressure gradient vanishes, when the pressure gradient at the Earth's surface, where $\bar{p} = 1000$ mb, is 0.011 mb km^{-1} and the temperature gradient at the same level 0.025 K km^{-1}? The pressure gradient and the temperature gradient are directed in opposite directions, and the horizontal temperature gradient should be considered as constant with height. The average temperature in the vertical is $\bar{T} = 272$ K. Treat the temperatures as virtual temperatures. (Hint: consider the layer thickness from the surface to the requested height.)

CHAPTER IX

VERTICAL STABILITY

In the previous chapter we have considered the atmosphere in hydrostatic equilibrium. It is obvious that this type of equilibrium does not imply a thermodynamic equilibrium; for instance, the vertical gradient of temperature will imply vertical heat conduction. In the real atmosphere heat will also be lost or gained by radiative processes. If vertical equilibrium is not prevalent, we should consider a third type of heat transport process: turbulent conduction. Conduction by molecular diffusion is a very slow process, negligible for all practical purposes, so that it need not concern us. Conduction by radiation is more important by several orders of magnitude, and it is a legitimate question to consider if radiative processes play a major role in determining the vertical distribution of temperature. Calculations for an atmospheric column which would receive radiative energy through the base, while losing the same amount from the top (admittedly a gross over-simplification of actual radiative processes), lead to a steady distribution such as indicated by the dashed curve of Figure IX-1. On the other hand, if we assume, in addition, a thorough vertical mixing, up to a level indicated by a discontinuity (tropopause), a distribution such as that of the full curve would be obtained. The real atmospheric stratifications are reasonably similar to this curve, especially for average conditions in temperate and tropical latitudes. The vertical turbulent transport of heat is thus a major factor

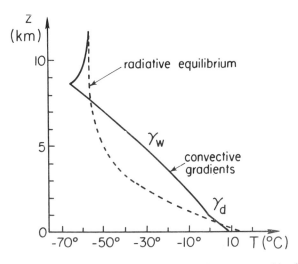

Fig. IX-1. Curve of radiative equilibrium and curve of convective lapse rates resulting from eddy diffusion (according to Emden).

determining the temperature distribution. In fact, when this mechanism is active, it may again be one or more orders of magnitude more efficient than the radiative transfer in determining local rates of change of temperature.

Apart from these considerations on energy transport, the vertical motions in the atmosphere are associated with weather disturbances and are therefore of primary importance to the meteorologist. Horizontal air motions can also not be neglected, and are particularly important on the time scale of day-to-day temperature changes.

In this chapter we shall investigate the conditions of vertical stability and instability in the atmosphere, first by assuming virtual infinitesimal displacements from equilirium at one point, then by looking into the consequences that finite vertical displacements of isolated air masses may have, and finally by considering vertical movements of extensive layers. We shall also consider the internal and potential energies of air columns, as well as the consequences of radiative processes.

In order to avoid confusion whenever this is liable to occur, we shall use different symbols for geometric and for process derivatives and differentials. In the first case we shall substitute the symbol δ for d (e.g.: writing δz, $\delta T/\delta \phi$, etc.), while we shall leave the usual d for the latter variations (e.g.: dz, $dT/d\phi$, etc.).

9.1. The Parcel Method

We shall now investigate the stability conditions regarding virtual vertical displacements of an atmospheric parcel in an environment which will be assumed to be in hydrostatic equilibrium.

The parcel is initially a part of the atmosphere, not different from any other at the same level. But it becomes an individualized portion as soon as it is assumed to start a displacement, while the environment remains at rest.

We shall denote by *reference level* the initial level of the parcel. The variables of the parcel will be distinguished by a prime (e.g., T') from those of the environment (T).

A number of simplifying assumptions are adopted in this method. It is supposed that:

(1) the parcel maintains its individuality during its movement, without mixing with the surrounding air,

(2) the movement of the parcel does not disturb the environment,

(3) the process is adiabatic, and

(4) at every instant the pressures of the parcel and of the environment for a given level are equal.

The first hypothesis, while reasonable when we consider infinitesimal virtual displacements, becomes quite unrealistic for finite displacements, and for this reason the data obtained in these cases are only semi-quantitative and must be corrected by empirical factors. The second assumption obviously cannot hold rigorously, since the ascent of an air mass must be compensated by the descent of other parts of the atmosphere; the approximation is good for isolated convection (when the disturbance of the environment is negligible) (cf. Section 8). The third hypothesis is reasonable, because the heat conduction processes in the atmosphere (turbulent diffusion, radiation, molecular conduction) are in general slow compared to convective movements. Pressure quickly attains equilibrium, and therefore the last hypothesis is also good, provided the motions are not so violent that hydrodynamic perturbations become appreciable.

Taking into account these considerations, it is clear that, although the method leads to correct criteria for vertical equilibrium when only infinitesimal virtual displacements are considered (with the restriction mentioned with respect to the second assumption), it will lead to quantitative values in considerable error when finite displacements are considered. Even in these cases, however, it sheds considerable light on significant aspects of vertical instability and gives correct qualitative conclusions.

9.2. Stability Criteria

Let us consider a parcel displaced from its initial position (reference level), and it environment at its new level. As the environment is in equilibrium, Chapter VIII, Equation (19) must hold:

$$\frac{\delta p}{\delta z} + g\varrho = 0. \tag{1}$$

The parcel, on the other hand, in general will not be in equilibrium: it will be subject to a force per unit mass equal to the resultant of the forces of gravity and of the pressure gradient. The first one is $-g$, where the sign indicates that it points downwards. To calculate the second one, we consider an infinitesimal layer of thickness δz from a column of unit cross section (see Figure IX-2). The force $-(\delta p/\delta z)\delta z = -(\delta p/\delta z)V'$ acts upon it, where V' is the volume of the layer. Dividing by V', we have the force per unit volume $-\delta p/\delta z$ (where p is the same as in Equation (1), according to the fourth assumption). The resultant of both forces will produce an acceleration $\ddot{z}$ (the dots meaning differentiation with respect to time), so that, per unit volume, we shall have the equation

$$\frac{\delta p}{\delta z} + g\varrho' = -\varrho'\ddot{z}. \tag{2}$$

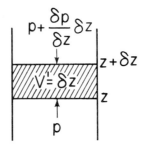

Fig. IX-2. A parcel, considered as a layer of an atmospheric column.

Subtracting Equation (2) from (1):

$$\ddot{z} = g\frac{\varrho - \varrho'}{\varrho'} \tag{3}$$

which, taking into account the gas law, and the definition of virtual temperature, can also be written

$$\ddot{z} = g\frac{v' - v}{v} = g\frac{T'_v - T_v}{T_v}. \tag{4}$$

This expression gives thus the force per unit mass acting on the parcel, due to both gravity and the pressure gradient. It is frequently called the *buoyancy* on the parcel, and written with the symbol B. In what follows we shall use the expression with virtual temperatures. We also have

$$T_v = T_{v_0} - \gamma_v\,\delta\phi\,; \qquad T'_v = T'_{v_0} - \gamma'_v\,d\phi' \tag{5}$$

where the subscript 0 refers to the reference level. Obviously, $T'_{v_0} = T_{v_0}$ and $\delta\phi = d\phi' = g\,dz$. Therefore:

$$T_v = T_{v_0} - \gamma_v g\,dz\,; \qquad T'_v = T_{v_0} - \gamma'_v g\,dz \tag{6}$$

and

$$T'_v - T_v = (\gamma_v - \gamma'_v)g\,dz\,. \tag{7}$$

Substituting into Equation (4):

$$\ddot{z} = \frac{g^2}{T_v}(\gamma_v - \gamma'_v)\,dz\,. \tag{8}$$

This formula shows that, if $\gamma'_v < \gamma_v$, the parcel will acquire an acceleration $\ddot{z}$ after the virtual displacement of the same sign as the displacement dz; it will tend to move away from the reference level, with increasing acceleration: this is the case of instability (positive buoyancy). If $\gamma'_v > \gamma_v$, $\ddot{z}$ and z will have opposite signs and the parcel will tend to return to its original position (negative buoyancy): this is the case of stability. If $\gamma'_v = \gamma_v$, the equilibrium is indifferent, and after being displaced, the parcel will remain in its new position. These stability conditions are summarized in the relation:

$$\gamma_v \gtreqless \gamma'_v \tag{9}$$

where
> corresponds to instability
= corresponds to indifference (neutral or zero stability)
< corresponds to stability.

The aerological sounding gives the variables T, p, U_w for each height; with these data γ_v can be calculated, and at the same time virtual displacements can be imagined and the resulting process lapse rate can be calculated. The comparison is extremely simple with a diagram, and permits the use of the criterion (9), as we shall see in Section 5.

9.3. Lapse Rates for Atmospheric Ascents

Before proceeding further with the stability criteria, we shall consider now the expressions for the thermal gradient for a dry adiabatic ascent γ_d, moist (air) adiabatic ascent γ_m, polytropic ascent γ_p and saturated adiabatic ascent γ_w. As the ascent of the parcel is adiabatic, the parcel will undergo one of these processes, according to the humidity conditions.

For an adiabatic process without condensation or evaporation:

$$\delta q = 0 = dh - v\, dp = c_p\, dT - v\, dp \tag{10}$$

and assuming hydrostatic equilibrium (i.e., assuming that the vertical motion is slow enough to consider the process as quasistatic or reversible), we use Chapter VIII, Equation (21):

$$c_p\, dT + d\phi = 0$$

$$\gamma_m = \frac{1}{c_p} \tag{11}$$

where $c_p \cong c_{pd}(1 + 0.87\, r)$ (see Chapter IV, Equation (87)).

In the particular case when the air is dry, $r = 0$ and

$$\begin{aligned}
\gamma_d = \frac{1}{c_{pd}} &= \frac{1}{1005}\text{ kg K J}^{-1} \\
&= 0.009\,76 \text{ K gpm}^{-1} \\
&= 9.76 \text{ K gpkm}^{-1} \\
&= 0.003\,20 \text{ K gpft}^{-1}.
\end{aligned} \tag{12}$$

This process lapse rate is equal to the geometric lapse rate of a dry adiabatic atmosphere (cf. Chapter VIII, Section 7).

The lapse rate for moist air will differ only slightly from γ_d, and can be expressed:

$$\gamma_m = \frac{\gamma_d}{1 + 0.87\, r} \cong \gamma_d(1 - 0.87\, r). \tag{13}$$

In the case of a polytropic ascent, as considered in Chapter VII, Section 11, δq is not zero in (10), and (11) is no longer valid. By following the same procedure as above, we derive in that case for the polytropic lapse rate γ_p:

$$\gamma_p = \frac{1}{c_p}\left(1 - \frac{\delta q}{d\phi}\right) = \frac{1}{c_p} - \frac{\lambda}{gU} \tag{14}$$

where λ and U have the same meanings as before. The second term becomes important only for extremely low rates of ascent ($U < 1$ cm s^{-1}).

The derivation of the lapse rate for saturated conditions is more laborious, and the result depends slightly on whether we assume a saturated reversible expansion (Chapter

VII, Section 8) or a pseudoadiabatic one (Chapter VII, Section 9). We shall begin with the first case. The starting point may be either Chapter VII, Equation (81), which expresses the condition $ds = 0$, or again

$$\delta q = dh - v\, dp = 0.$$

Both are equivalent, for the process is assumed to be reversible, and therefore $\delta q = T\, ds$. We shall have

$$\gamma_w = -\frac{dT}{d\phi} = \frac{1}{v}\frac{dT}{dp} = \frac{p}{RT}\frac{dT}{dp} \tag{15}$$

where $R = R_d(1 + 0.61 r_w)$. The expression for dT/dp to be introduced here is found from the equation describing the process. We write Chapter VII, Equation (81), developing the differentials:

$$\frac{c_{p_d} + c_w r_{t,w}}{T} dT - \frac{R_d}{p_d} dp_d + \frac{l_v}{T} dr_w + \frac{r_w}{T} dl_v - \frac{r_w l_v}{T^2} dT = 0. \tag{16}$$

We use now the substitutions

$$p = p_d + e_w \tag{17}$$

$$dp_d = dp - de_w \tag{18}$$

$$r_w = \frac{\varepsilon e_w}{p - e_w} \tag{19}$$

$$dr_w = \frac{\varepsilon\, de_w}{p - e_w} - \frac{\varepsilon e_w\, dp}{(p - e_w)^2} + \frac{\varepsilon e_w\, de_w}{(p - e_w)^2}$$

$$= \frac{\varepsilon + r_w}{p - e_w} de_w - \frac{\varepsilon e_w}{(p - e_w)^2} dp, \tag{20}$$

we divide Equation (16) by dp and we further substitute

$$\frac{de_w}{dp} = \frac{de_w}{dT}\frac{dT}{dp} \tag{21}$$

$$\frac{dl_v}{dp} = \frac{dl_v}{dT}\frac{dT}{dp}. \tag{22}$$

We then solve for dT/dp, obtaining finally:

$$\frac{dT}{dp} = \frac{\dfrac{R_d}{p - e_w} + \dfrac{\varepsilon l_v e_w}{T(p - e_w)^2}}{\dfrac{c_{p_d} + c_w r_{t,w}}{T} - \dfrac{l_v r_w}{T^2} + \dfrac{r_w}{T}\dfrac{dl_v}{dT} + \dfrac{de_w}{dT}\left[\dfrac{R_d}{p - e_w} + \dfrac{l_v(\varepsilon + r_w)}{T(p - e_w)}\right]}. \tag{23}$$

If we introduce this expression into expression (15), take into account the relation between R and R_d, Equations (12), (19), and the Kirchhoff and Clausius-Clapeyron equations:

$$\frac{dl_v}{dT} = c_{p_v} - c_w \tag{24}$$

$$\frac{de_w}{dT} = \frac{l_v e_w}{R_v T^2} = \frac{\varepsilon l_v e_w}{R_d T^2} \tag{25}$$

we can obtain the expression

$$\gamma_w = \gamma_d \frac{\dfrac{p}{p - e_w}\left[1 + \left(\dfrac{l_v}{RT} - 0.61\right)r_w\right]}{1 + \dfrac{c_{p_v} r_w + c_w(r_{t,w} - r_w)}{c_{p_d}} + \dfrac{l_v^2 r_w(\varepsilon + r_w)}{c_{p_d} R_d T^2}}. \tag{26}$$

This formula gives the lapse rate for the reversible saturated adiabatic expansion, and depends slightly on the proportion of liquid water referred to the unit mass of dry air $(r_{t,w} - r_w)$. The formula for the pseudoadiabatic process will be obtained from Equation (26) by setting $(r_{t,w} - r_w) = 0$, i.e., by suppressing in the denominator the term with c_w.

Equation (26) can be simplified by several approximations. In the numerator, $0.61 \ll l_v/RT$, so that the term 0.61 can be neglected. The ratio $p/(p - e_w)$ can differ by a few per cent from unity; it can be set equal to unity within that approximation. The second term in the denominator can also amount to a few per cent, and if we neglect it, the error will be partly compensated by the previous approximation. Finally $r_w \ll \varepsilon$ so that r_w can also be neglected in the last term of the denominator. The result of these simplifications is the formula

$$\gamma_w \cong \gamma_d \frac{1 + \dfrac{l_v r_w}{RT}}{1 + \dfrac{\varepsilon l_v^2 r_w}{c_{p_d} R_d T^2}}. \tag{27}$$

And writing $r_w \cong \varepsilon e_w/p$, $R \cong R_d$,

$$\gamma_w \cong \gamma_d \frac{1 + \dfrac{\varepsilon l_v e_w}{R_d T p}}{1 + \dfrac{\varepsilon^2 l_v^2 e_w}{c_{p_d} R_d T^2 p}}. \tag{28}$$

This approximate expression does not distinguish between the reversible and the pseudoadiabatic process. It might have been derived directly from the approximate expression Chapter VII, Equation (83) with $c_p \cong c_{pd}$ by following a similar procedure to that leading to expression (26), introducing the same approximations as for the derivation of Equation (28) and making the additional simplification of neglecting unity against $\varepsilon l_v / R_d T \sim 20$.

We can illustrate the errors involved in these approximations by giving a numerical example. Let us assume that $T = 17\,°C$, $p = 1000$ mb; thus $e_w = 19.4$ mb, $r_w = 0.0123$. Assuming that $r_{t,w} = 0.0163$ (corresponding to a large liquid water content of 4 g kg^{-1} of dry air), Equation (26) gives

$$(\gamma_w)_{rev} = 4.40 \text{ K gpkm}^{-1}.$$

Setting $r_{t,w} = r_w$ (no liquid water), Equation (26) gives for the pseudoadiabatic process:

$$(\gamma_w)_{ps} = 4.42 \text{ K gpkm}^{-1}$$

i.e., a value only 0.5% higher. The approximate formula (28) gives

$$\gamma_w \cong 4.53 \text{ K gpkm}^{-1}.$$

With numerical values of the constants, Equation (28) reads

$$\gamma_w \cong 9.76 \, \frac{1 + 5.42 \times 10^3 \, e_w / Tp}{1 + 8.39 \times 10^6 \, e_w / T^2 p} \text{ K gpkm}^{-1}. \tag{29}$$

Equation (26), or the approximate relations (28) or (29), permits the computation of γ_w as a function of T and p. The second term in the numerator of Equation (28) is always smaller than the second term in the denominator (because $\varepsilon < l_v / c_{pd} T = = 2500/T$ in the atmosphere), and therefore $\gamma_w < \gamma_d$. The largest differences are found for high values of T (and therefore high values of e_w), when γ_w decreases to near 3 K gpkm^{-1}. For lower T, e_w decreases rapidly and, unless p is also very low, the second terms in the numerator and denominator become negligible, so that γ_w tends to the value $\gamma_d = 9.76$ K gpkm^{-1}. In an aerological diagram this means that for decreasing temperatures the saturated adiabats tend to coincide with dry adiabats of the same pseudo-equivalent potential temperature θ_{ae}.

The previous derivations refer to water clouds. The same formulas hold for ice clouds, provided we make the corresponding substitutions: l_s, e_i and r_i for l_v, e_w and r_w.

9.4. The Lapse Rates of the Parcel and of the Environment

Let us go back now to the stability criteria (9) and consider the virtual temperature lapse rates γ'_v and γ_v.

According to the initial assumptions, the virtual processes undergone by the parcel must be adiabatic. If the parcel is not saturated, as there is no mixing with the surroundings, the mixing ratio r_0 will be a constant, and the process is a moist adiabatic expansion (or compression). By differentiating

$$T'_v = (1 + 0.61 r_0) T' \tag{30}$$

with respect to ϕ and changing sign, we obtain:

$$\gamma'_v = (1 + 0.61\, r_0)\gamma_m = (1 + 0.61\, r_0)(1 - 0.87\, r_0)\gamma_d = (1 - 0.26\, r_0)\gamma_d \quad (31)$$

where γ_m is the lapse rate for the moist adiabat and γ_d that for the dry adiabat.

The term $0.26\, r_0$ is small compared with unity and therefore we may neglect it as a first approximation. Thus

$$\gamma'_v \cong \gamma_d \quad (unsaturated\ parcel). \quad (32)$$

If on the other hand the parcel is saturated, its mixing ratio r_w must decrease during the ascent. By proceeding as before, we have:

$$T'_v = (1 + 0.61\, r'_w)T' \quad (33)$$

$$\gamma'_v = (1 + 0.61\, r'_w)\gamma_w - 0.61\, T' \frac{dr'_w}{d\phi} \quad (34)$$

where γ_w is the saturated adiabatic lapse rate.

We remark now that $dr'_w/d\phi < 0$, $\gamma'_v > \gamma_w$, and as r'_w and its derivative decrease with ascent, γ'_v becomes closer to γ_w. For the lower troposphere the last term may amount to as much as 10% of γ_w; if we neglect this term as a first approximation, as well as $0.61\, r_w$ against unity,

$$\gamma'_v \cong \gamma_w \quad (saturated\ parcel). \quad (35)$$

Therefore, we identify the lapse rate of the parcel approximately with γ_d or γ_w, according to whether it is unsaturated or saturated. In Section 3 we have derived the values for these two quantities.

We may now consider the lapse rate of the environment in a similar way, but remembering that here we are dealing with a geometric derivative rather than with a physical process. We shall have

$$T_v = (1 + 0.61\, r)T \quad (36)$$

$$\gamma_v = (1 + 0.61\, r)\gamma - 0.61\, T \frac{\delta r}{\delta \phi}. \quad (37)$$

If we take into account the relation (9), the last term indicates that according to whether $\delta r/\delta\phi \gtreqless 0$, the humidity distribution with height will increase or decrease the vertical stability. In certain conditions this term may become quite appreciable. Let us consider an example. We assume $\gamma = 0$, for simplicity. Now we ask what distribution or r may determine a value $\gamma_v = l/c_{p_d} = \gamma_d$ for the lapse rate. It will be:

$$\frac{\delta r}{\delta \phi} = -\frac{\gamma_d}{0.61\, T} \cong -6 \times 10^{-5}\ \text{gpm}^{-1}.$$

This condition would be satisfied, for instance, if the mixing ratio decreased by

8.6×10^{-3} along 150 gpm. This variation could be present in a layer whose base was saturated at 10 °C and 900 mb and whose top was dry. Such vertical humidity gradients could arise in the case of dry inversion layers above cloud strata.

We may also define a potential temperature lapse rate and derive its value for an unsaturated atmosphere as a function of the lapse rate γ_v. If we differentiate logarithmically the definition of θ_v:

$$\theta_v = T_v \left(\frac{1000}{p}\right)^{\varkappa_d} \quad (p \text{ in mb}), \tag{38}$$

we obtain

$$\frac{1}{\theta_v} \frac{\delta \theta_v}{\delta \phi} = \frac{1}{T_v} \frac{\delta T_v}{\delta \phi} - \frac{\varkappa_d}{p} \frac{\delta p}{\delta \phi}. \tag{39}$$

If we now introduce Chapter VIII, Equation (21), the definition of $\varkappa_d$, the gas law and the value of γ_d, we obtain

$$\frac{\delta \theta_v}{\delta \phi} = \frac{\theta_v}{T_v} (\gamma_d - \gamma_v) \tag{40}$$

which gives, except for the sign, the lapse rate of θ_v. This is therefore proportional to the difference between γ_v and the adiabatic lapse rate. If $\gamma_v = \gamma_d$, we have $\theta_v = \text{const.}$ (adiabatic atmosphere).

9.5. Stability Criteria for Adiabatic Processes

Within the approximations made in the previous paragraph, we may now write the criteria (9) in the following form:

$$\gamma_v \gtreqless \gamma_d \quad \textit{unsaturated parcel}, \tag{41}$$

$$\gamma_v \gtreqless \gamma_w \quad \textit{saturated parcel}, \tag{42}$$

where, as before, the sign $>$ corresponds to *instability*, the sign $=$ to *indifference*, and the sign $<$ to *stability*.

We have here relations between thermal gradients, that is between derivatives of the temperature with respect to ϕ. The geopotential is proportional to the height z, and the variables $-p$, $-\ln p$ or $-p^\varkappa$ increase monotonically with z. Therefore, the relative values of the derivatives T with respect to these variables will follow the same order as the lapse rates γ. In the diagrams (tephigram, emagram, Stüve) where T is the abscissa, the curves giving the variation of T will appear more inclined backwards the higher the value of γ. For any of these representations we may thus visualize the conditions (41) and (42) by means of the diagram shown in Figure IX-3.

The procedure consists in observing, for a point P_0 of the state curve, what relation the slope of the curve bears to the slopes of the dry and saturated adiabats. According to the slope, the curve will fall within one of the stability regions; thus

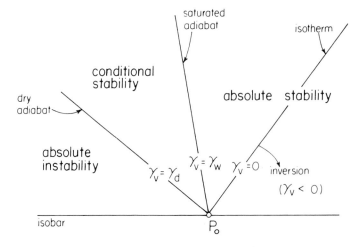

Fig. IX-3. Lapse-rate regions of stability, on a diagram.

the zone of absolute instability corresponds to $\quad \gamma_v > \gamma_d$

the zone of conditional instability corresponds to $\quad \gamma_w < \gamma_v < \gamma_d$

the zone of absolute stability corresponds to $\quad \gamma_v < \gamma_w$

The word 'absolute' indicates that the stability or instability condition holds independently of the air saturation. The designation of "conditional instability" means that, if the air is saturated, there will be instability ($\gamma'_v \cong \gamma_w < \gamma_v$), and in the opposite case, there will be stability ($\gamma'_v \cong \gamma_d > \gamma_v$).

Let us consider now an unsaturated parcel in the case of instability. We shall have $\gamma_v > \gamma_d$, and in a diagram the state curve (γ_v) and the process curves (γ_d) will appear as in Figure IX-4, where θ, θ', θ'', θ''' indicate dry adiabats of decreasing potential temperatures. It is obvious that for rising ϕ, the state curve cuts dry adiabats of

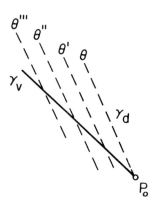

Fig. IX-4. Potential temperature decrease with height for a superadiabatic lapse rate.

decreasing potential temperature θ; that is, $\delta\theta_v/\delta\phi<0$. If we had started with the stability case, we would have arrived at the opposite result. On the other hand, the argument may be applied in a similar way to the case of a saturated parcel, if instead of γ_d we consider the lapse rate γ_w, and we substitute the potential wet bulb temperature θ_{aw} or the potential equivalent temperature θ_{ae} for θ, these temperatures being the

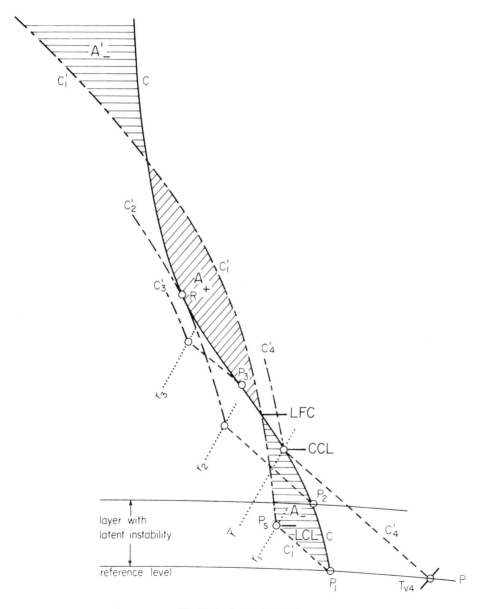

Fig. IX-5. Latent instability.

invariants characteristic of the saturated adiabats. We may therefore express the stability conditions in the following alternative way:

$$\frac{\delta \theta_v}{\delta \phi} \lessgtr 0 \qquad \textit{unsaturated parcel} \qquad (43)$$

$$\frac{\delta \theta_{aw}}{\delta \phi} \lessgtr 0 \quad \text{or} \quad \frac{\delta \theta_{ae}}{\delta \phi} \lessgtr 0 \quad \textit{saturated parcel} \qquad (44)$$

where now the sign < corresponds to *instability*, the sign = to *indifference*, and the sign > to *stability*. Obviously the same conditions can be written with the z derivatives, instead of using the geopotential.

The relation (43) could also have been derived by introducing Equation (41) in (40).

9.6. Conditional Instability

In Section 5 we have denoted by conditional instability the case $\gamma_w < \gamma_v < \gamma_d$. All arguments so far have been based on infinitesimal virtual displacements. They have been used therefore to analyze the stability conditions at a certain point or, as we are studying an atmosphere where all properties are constant on equipotential surfaces, at a given level.

We shall now study the stability conditions in a conditionally unstable layer, such as that extending from the surface (P_1) to the point R of the state curve in Figure IX-5 (representing a tephigram). All the points of the state curve within that layer obey the conditional instability criterion. But we shall now discuss what happens when parcels coming from different levels rise along the vertical, in finite displacements.

The temperature distribution of the atmosphere is represented by the curve c (sounding). The curves c'_1, c'_2 and c'_3 represent the processes undergone by parcels coming from the reference levels P'_1, P'_2 and P'_3, respectively.

When a parcel rises vertically in the atmosphere, a certain amount of work is performed by or against the buoyancy forces according to whether the process curve lies at the right or at the left of the state curve. We shall now show that this work is proportional to the area enclosed between the two curves and the isobars defining the initial and final levels, in any area-preserving diagram.

Let us first consider the process on an emagram (Figure IX-6). The work done by the buoyant force on the parcel, per unit mass, will be

$$w = \int_a^b \ddot{z} \, dz, \qquad (45)$$

and taking into account Equation (4), (Chapter VIII, Equation (21)) and the gas law,

$$w = g \int_a^b \frac{v' - v}{v} \, dz = -\int_a^b (v' - v) \, dp$$

$$= R_d \int_a^b (T'_v - T_v) \, d(-\ln p)$$

$$= R_d (\Sigma'_{em} - \Sigma_{em}), \tag{46}$$

where Σ'_{em} and Σ_{em} are the areas defined by the two isobars, the vertical axis, and the curve c' or c, respectively. The difference is the area shaded in the figure. This is a closed area which might be considered as the representation of a cycle. We have seen in Chapter VI that the same cycle will define a proportional area in any other area-preserving diagram, such as, for instance, the tephigram.

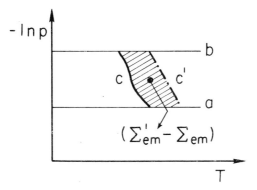

Fig. IX-6. Work performed on a parcel by buoyancy.

Let us go back now to the process that we are considering. The work received per unit mass of the parcel will be transformed into kinetic energy:

$$w = \int_a^b \ddot{z} \, dz = \int_a^b \frac{d\dot{z}}{dt} \, dz = \int_a^b \dot{z} \, d\dot{z} = \tfrac{1}{2}(\dot{z}_b^2 - \dot{z}_a^2), \tag{47}$$

where $\tfrac{1}{2}\dot{z}^2$ is the kinetic energy of the unit mass. If c' is to the right of c, as in the figure, w is positive: work is done on the parcel by the external forces of gravity and pressure, and the parcel accelerates. If c' is to the left of c, w is negative and the parcel decelerates; in order to rise, the parcel must be provided with an energy equal to the negative area that it determines on the diagram along the ascent.

In Figure IX-5, in order for the parcel to follow the path c'_1, work proportional to the area A_- must be performed against the negative buoyant force (for instance, the necessary energy could come from a forced orographic ascent). The level at which it reaches saturation (point P_s) is called the *lifting condensation level* (LCL).

When the parcel surpasses the level at which the process curve crosses the state curve, the buoyant force performs a work proportional to the area A_+, up to the isobar reached. Convection will thus continue freely until c_1' crosses again c, and the parcel decelerates. The level of the first crossing is called the *level of free convection* (LFC). The total area A_+ measures the *latent instability* for a parcel at the reference level P_1.

One can proceed similarly with parcels from other levels. When the level of P_2 is reached, A_+ vanishes: this is the upper limit of the layer for which there is latent instability. For instance, c_3' does not show any latent instability.

It may be noticed that the instability is greater for higher T and r. When r increases, the segment $P_1 P_s$ becomes shorter, A_- decreases and A_+ increases. Regarding the temperature, if we assume that after the sounding c, the ground is warmed up by radiation (insolation), a steeper lapse rate γ_v will appear in the lowest layer; when it exceeds the dry adiabatic lapse γ_d, the layer becomes unstable and a vertical mixing process starts working, which results in a dry adiabatic lapse rate for that layer (cf. Chapter VII, Section 12). As the warming continues, an increasing depth of the lowest part of the sounding becomes substituted by dry adiabats of increasing potential temperature. Eventually, the top of the stirred layer may reach saturation, that is, the dry adiabatic portion c_4' reaches the mixing ratio isopleth corresponding to the mean value for the layer $\bar{r}$. This level is called the *convective condensation level* (CCL). The temperature of the ground has reached then the value T_{v_4} (see Figure IX-5) and from that moment, convection may proceed spontaneously (along c_4', dry adiabat up to CCL and then saturated adiabat) without any need of forced lifting.

We have mentioned that in Figure IX-5 the lower part of the sounding has been assumed to have lapse rates between γ_d and γ_w, corresponding to conditional instability. Conditional instability is sometimes classified into three types:

Conditional instability
- of the real *latent type*: when $A_+ > A_-$, as in Figure IX-5.
- of the *pseudolatent type*: when $A_+ < A_-$.
- of the *stable* type: when no positive area A_+ appears for parcels of any level.

Because a measure of the instability area can take too long for routine application, it is frequently substituted by the determination of the difference of temperature $T_v - T_v'$ at 500 mb for parcels rising from the 850 mb level; this value is called the *Showalter stability index*. The larger positive this index is, the greater is the local stability. Large negative areas, on the other hand, will correspond to large negative values (i.e., several degrees) of the index.

It must be kept in mind that this analysis of vertical stability must be assessed in the context of the synoptic situation and taking into account the approximations involved. For instance, the difference of areas $A_+ - A_-$ may not be particularly important if A_- is large enough to prevent the onsetting of convection. Conversely, small negative areas may easily be eliminated during insolation or may not even be meaningful, in view of the local temperatures' contrasts, which are not taken into account in the analysis; these differences will exist owing to the different characteristics of the ground surface (low heat capacity and conductivity of the ground will favor overheating of its surface) and convection (updrafts or thermals) will start over the areas at higher temperature.

9.7. Oscillations in a Stable Layer

If some disturbance provokes a vertical displacement in a stable layer ($\gamma'_v < \gamma_v$), an oscillatory motion results. This will actually be damped by turbulent mixing of the borders of the parcel with the environment. For the parcel method, which does not take into account this effect, the motion will be undamped.

The acceleration is (formula (8)):

$$\ddot{z} = -\frac{g^2}{T_v}(\gamma'_v - \gamma_v)z \qquad (48)$$

where z is now written for the displacement. This is the equation for the motion of a linear harmonic oscillator:

$$\ddot{z} + kz = 0, \qquad (49)$$

where the force constant is

$$k = \frac{g^2}{T_v}(\gamma'_v - \gamma_v) = \omega^2 = \left(\frac{2\pi}{\tau}\right)^2 \qquad (50)$$

(ω = angular frequency; τ = period). The solution is

$$z = A \sin \omega t, \qquad (51)$$

and the period

$$\tau = \frac{2\pi}{g}\sqrt{\frac{T_v}{\gamma'_v - \gamma_v}} \qquad (52)$$

will be larger, the smaller the difference between the lapse rate of parcel and environment. The angular frequency of such (gravity) waves is usually referred to as the Brunt-Vaisala frequency.

For example, let us assume that

$$T_v = 0\,°C, \qquad \gamma_v = 0, \qquad \gamma'_v \cong \gamma_d = 1/c_{p_d} \quad \text{(unsaturated parcel)}.$$

We obtain:

$$\tau = 335 \text{ s} = 5.6 \text{ min}$$

If $A = 200$ m, the maximum velocity of the oscillation will be

$$\dot{z}_{max} = A\omega = \frac{200 \times 2\pi}{335} = 3.7 \text{ m s}^{-1}.$$

9.8. The Layer Method for Analyzing Stability

As we remarked in Section 1, one of the main defects of the parcel method is the assumption that the environment remains undisturbed. The ascent of an air mass must necessarily be compensated by the descent of surrounding air. If the rising masses cover an appreciable fraction of the total area, the error becomes important. In order to allow for the compensatory descending motion, Bjerknes devised the layer, or 'slice', method, which we shall describe now.

Let us consider a certain level, over an area large enough to cover a representative number of possible ascending currents (Figure IX-7). Let us discuss a virtual process by which updraughts start, covering an area A', while this movement is compensated

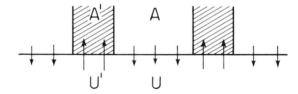

Fig. IX-7. Ascending and descending motions according to the layer method.

by a general descent of the environment, covering an area A. We assume that the velocities are constant and equal to U' (upwards) and U (downwards). The layer is initially uniform. The total mass of ascending and descending air must be equal:

$$A'U' = AU. \tag{53}$$

For a time interval dt, the displacements of ascending and descending air will be dz' and dz, respectively, and

$$\frac{U}{U'} = \frac{dz/dt}{dz'/dt} = \frac{dz}{dz'} = \frac{A'}{A} \tag{54}$$

where the velocities and displacements are expressed as absolute values.

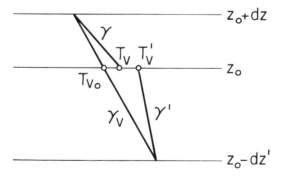

Fig. IX-8. Layer method. Case $\gamma_w < \gamma_v < \gamma_d$ with instability.

The stability conditions will result, as in the parcel method, from the value of the acceleration $\ddot{z}$ of the rising air as given by Equation (4). In order to calculate the difference $(T'_v - T_v)$, we shall consider the level z_0 of the layer at the time dt of the virtual displacement. The air crossing z_0 upwards comes from the level $z_0 - dz'$, and the air crossing downwards, from $z_0 + dz$ (Figure IX-8). γ_v, γ' and γ are, respectively, the geometric lapse rate of the atmosphere at z_0 (at the initial time), and the process lapse rate (for virtual temperature, although the subscript v has been dropped here for convenience) of ascending and descending air (approximately γ_d or γ_w, as the case may be, according to Section 4). If the initial (virtual) temperature at z_0 is T_{v_0}, at the time dt we shall have, for the temperatures of the ascending and descending air:

$$T'_v = T_{v_0} + \gamma_v g\, dz' - \gamma' g\, dz' = T_{v_0} + (\gamma_v - \gamma') g\, dz'$$

$$T_v = T_{v_0} - \gamma_v g\, dz + \gamma g\, dz = T_{v_0} + (\gamma - \gamma_v) g\, dz$$

$$\Delta T_v = T'_v - T_v = (\gamma_v - \gamma') g\, dz' - (\gamma - \gamma_v) g\, dz$$

$$= g\left[(\gamma_v - \gamma') - \frac{A'}{A}(\gamma - \gamma_v)\right] dz' \tag{55}$$

where the relation (54) has been introduced. Replacing ΔT_v in Equation (4) by (55), we have for the rising air:

$$\ddot{z} = \frac{g^2}{T_v}\left[(\gamma_v - \gamma') - \frac{A'}{A}(\gamma - \gamma_v)\right] dz',$$

and arguing as in Section 2, we obtain as the stability criterion, that in the relation

$$(\gamma_v - \gamma') - \frac{A'}{A}(\gamma - \gamma_v) \gtreqless 0 \tag{56}$$

the sign $>$ corresponds to *instability*, the sign $=$ to *indifference*, and the sign $<$ to *stability*.

It may be noticed that for the particular case in which the area of rising air is a negligible fraction of the total area, $A'/A \cong 0$, and Equation (56) becomes equal to the conditions (9) derived with the parcel method.

Let us consider now the different possible cases.

(I) $\gamma_w < \gamma_v < \gamma_d$ (conditional instability) and the air is saturated at the level z_0. It is the case of a layer at the condensation level. The rising saturated air will follow a saturated adiabat, while the descending air will follow a moist (approximately a dry) adiabat: $\gamma' \cong \gamma_w$ and $\gamma \cong \gamma_d$. Condition (56) becomes

$$\frac{\gamma_v - \gamma_w}{\gamma_d - \gamma_v} \gtreqless \frac{A'}{A} = \frac{U}{U'} \tag{57}$$

where the upper sign corresponds always to instability. Stability does not depend only on the value of γ_v, but also on the relative extent of convection. Instability will be

194   ATMOSPHERIC THERMODYNAMICS

reached more easily if this extent is small (i.e., small $A'/A = U/U'$). Figure IX-8 corresponds to instability, and Figure IX-9 to stability.

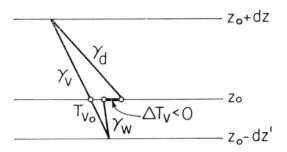

Fig. IX-9.   Layer method. Case $\gamma_w < \gamma_v < \gamma_d$ with stability.

(II) $\gamma_v > \gamma, \gamma'$. Whatever may be the values of γ, γ' and A'/A, Equation (56) indicates instability, because both terms on the left side are positive. Let us notice that, according to this method, ΔT_v is larger than for the parcel method, which only gives the first term of Equation (55).

This may be the case of unsaturated air rising in an unsaturated environment ($\gamma = \gamma' = \gamma_d$), of saturated air in a saturated environment ($\gamma = \gamma' = \gamma_w$), or of saturated air rising in an unsaturated environment ($\gamma = \gamma_d$; $\gamma' = \gamma_w$); in the last case the reference level is the saturation level.

These cases are shown by Figure IX-10.

(III) $\gamma_v < \gamma, \gamma'$. Whatever may be the values of γ, γ' and A'/A, Equation (56) indicates stability, because both terms on the left are negative. Again the difference ΔT_v (negative) is larger in absolute value than in the parcel method.

Figure IX-11 corresponds to this case, for which we have the same possibilities as in (II) regarding saturation.

Thus, the layer method shows us that the effect of the environment subsidence due to the convection is as follows: when the parcel method indicates absolute instability, this effect makes the instability even more pronounced; when there is absolute stability, it makes it even more stable; when there is conditional instability, the environment

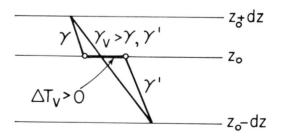

Fig. IX-10.   Layer method. Case $\gamma_v > \gamma, \gamma'$ (absolute instability).

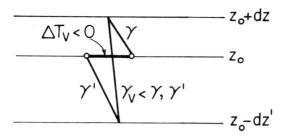

Fig. IX-11. Layer method. Case $\gamma_v < \gamma, \gamma'$ (absolute stability).

subsidence qualifies it in such a way that the instability becomes less pronounced and may even revert to stability, according to the extent of the convection and to the particular value of γ_v.

The layer method constitutes an improvement over the parcel method in that it disposes of one of its most objectionable assumptions. It provides a satisfactory tool for the analysis of stability at a given level by considering infinitesimal virtual motions. Its application, however, requires an estimate of the parameter A'/A, and this would generally be an arbitrary guess.

9.9. Entrainment

When the parcel method is applied to finite vertical displacements, as in the analysis of conditional instability (Section 6), the assumption of no mixing with the surroundings remains as the main source of error; in fact, turbulent mixing is very active, and ignoring it is the cause of obtaining exaggerated values for the temperature difference between parcel and environment, the kinetic energy acquired by a saturated parcel and its liquid water content. The main effect of mixing is to incorporate into the rising parcel a certain amount of external air that becomes mixed with the rest of the parcel; this is called 'entrainment'. There is also the possibility of 'detrainment', i.e., of a certain proportion of the rising parcel being shed and coming to a halt while mixing with the surrounding air. If we take into account only the first effect and an estimate can be made for the proportion of external air being entrained per unit length of ascent (rate of entrainment), a convenient correction of the parcel method predictions can be worked out. This will now be explained. The rate of entrainment, however, is a highly variable parameter, depending on the stage of development of the cloud, the dimensions of the ascending mass and the intensity of the convection (vertical velocities), therefore difficult to estimate, but the method will show how the effect of entrainment on convective parameters can be visualized by a simple graphical procedure on a diagram.

We divide the ascent into a number of steps, and consider each step as consisting of three processes:

(a) the air rises $\varDelta z$, without mixing, as in the parcel method;

(b) it mixes isobarically with a certain proportion of the surrounding air (cf. Chapter VII, Section 4); and

(c) liquid water from the parcel evaporates, until the gaseous phase becomes again saturated (adiabatic isobaric evaporation), or until the water has completely evaporated. This process will not be required if the parcel is not saturated before or after mixing.

By repeating this procedure in successive steps, a much more realistic curve can be obtained for the rising parcel properties whenever we can have a reasonable estimation of the proportion of external air incorporated in each step.

The correction could be computed from the formulas in Chapter VII, but it is more conveniently estimated on a diagram. Let us consider Figure IX-12, where one step

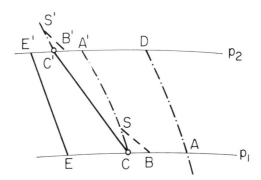

Fig. IX-12. Graphical correction for entrainment.

of the process is followed on a tephigram. We assume that EE' is the state curve of the environment, AD the saturated adiabat that the parcel would follow if it ascended without mixing from the point A (reached from a previous step, process (a)). We shall assume that in every step 50% of external air is incorporated (and no air from the parcel is lost). At the level p_1 isobaric mixing takes place (process (b)); the point B results from mixing air masses E and A in the assumed proportions. According to the rule of mixtures, the segment BA will be in our example one-third of EA, because the resulting temperature must be the weighted average of T_E and T_A. Process (c) is substituted with sufficient approximation by the process defining the pseudo-wet-bulb temperature (Chapter VII, Section 13): ascent along the dry adiabat until saturation (point S), and descent along the saturated adiabat to level p_1. We obtain thus point C, representing the parcel at level p_1.

A new ascent to the next step will bring the parcel to A' at level p_2, where the processes of mixing and evaporation are performed again in a similar way. Point C' is then obtained. The segment CC' is the corrected trajectory of the parcel between the levels p_1 and p_2. By repeating the procedure, new segments will be added, and the whole curve, between the convective condensation level and the level at which all the liquid water disappears after crossing the state curve to the left, can be constructed.

It will be remarked that in each step the resulting humidity of the isobaric mixture has to be computed analytically in order to be able to determine the saturation point S.

Entrainment can have a major effect on cumulus convection processes, and in particular on their evolution in time. If the environment air surrounding the cumulus cloud is relatively dry, as it will be during the early stages of the growth of the cloud, detrainment (or external entrainment) will result in dissolution of cloud elements, leaving, however, a moister environment than previously existed. The temperature of the rising air will decrease as a result of the initial mixing and of the subsequent evaporation, reducing or even nullifying the relative buoyancy of the rising parcel. The same process will cause isolated cloud turrets to dissolve and appear to fall back into the parent cloud. Internal entrainment will similarly reduce the liquid water content and buoyancy of cloud elements inside the cloud, so that the net effect of these processes is to produce horizontal gradients of liquid water content and vertical velocities, with maximum values in the interior of the cloud. Only in the central core of a cumulus cloud can one achieve theoretical values of liquid water content (usually known as 'full adiabatic' values). Vertical velocities, even at the core, seldom reach their theoretical values because of the reduction of cloud buoyancy, and as a result cumulus cloud tops seldom reach the predicted level (of equilibrium with the environment). A further consequence of these mixing processes is that the cloud soon develops a dome shape, rather than a cylindrical shape. If an ample supply of energy exists, the damping effect of the environment will decrease as its relative humidity increases, until eventually a more-nearly columnar development of the cloud to high levels is possible and the cumulus cloud becomes a cumulus congestus.

9.10. Potential or Convective Instability

With the parcel method, we have so far considered the stability properties of the atmosphere when an isolated mass, the parcel, is vertically displaced. This occurs, for instance, when the warming of the lower layers causes the ascent of air masses with dimensions of the order of hundreds of meters to 1 km. These masses can eventually become accelerated by a latent instability, as we have seen. They are called *thermals* or *bubbles*, and become visible as convective clouds. When the convection becomes more intense we may have, instead of isolated masses, a continuous jet of ascending air. The previous calculations would apply equally to this type of convection.

It is also important to study the vertical movements of an extended layer of atmosphere, such as may occur, for instance, during the forced ascent of an air mass over an orographic obstacle, or due to large-scale vertical motions of appreciable magnitude (as, for example, in frontal situations). We shall now consider this case, deriving the effect of the movement on the lapse rate of the layer.

9.10.1. THE LAYER IS AND REMAINS NON-SATURATED

We represent the layer δz in Figure IX-13, where the right hand side is a diagram z, T (the relations would be similar on a tephigram). This layer ascends from p to p'; at the first level the area considered will be A' and the thickness $\delta z'$. The virtual

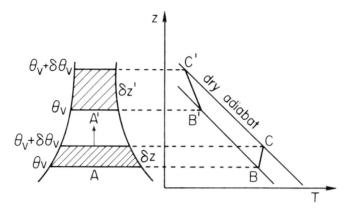

Fig. IX-13. Potential instability, non-saturated layer.

potential temperature θ_v of the base of the layer will remain constant during the ascent, which we assume adiabatic, and the same will happen with the temperature $\theta_v + \delta\theta_v$ of the top of the layer; therefore the variation $\delta\theta_v$ through the layer also remains constant.

Conservation of the total layer mass requires that

$$\varrho A \, \delta z = \varrho' A' \, \delta z' \tag{58}$$

or

$$\frac{\delta z'}{\delta z} = \frac{\varrho A}{\varrho' A'} = \frac{pAT'_v}{p'A'T_v} \tag{59}$$

where all the primed variables correspond to the final state, the gas law having been introduced in the last expression.

As $\delta\theta_v$ remains constant, the gradient of θ_v must be inversely proportional to δz. We must have:

$$\frac{\delta\theta_v}{\delta z} = \frac{\delta\theta_v}{\delta z'} \frac{\delta z'}{\delta z} = \frac{\delta\theta_v}{\delta z'} \frac{pAT'_v}{p'A'T_v}. \tag{60}$$

Or, dividing both sides by g:

$$\frac{\delta\theta_v}{\delta\phi} = \frac{\delta\theta_v}{\delta\phi'} \frac{pAT'_v}{p'A'T_v}, \tag{61}$$

and introducing Equation (40):

$$\frac{\theta_v}{T_v}(\gamma_d - \gamma_v) = \frac{\theta_v}{T'_v}(\gamma_d - \gamma'_v) \frac{pAT'_v}{p'A'T_v} \tag{62}$$

and solving for γ'_v:

$$\gamma'_v = \gamma_d - \frac{p'A'}{pA}(\gamma_d - \gamma_v)$$

$$= \gamma_v + (\gamma_d - \gamma_v)\left(1 - \frac{p'A'}{pA}\right). \tag{63}$$

If $\gamma_v = \gamma_d$, $\gamma'_v = \gamma_d = \gamma_v$; that is, the layer is adiabatic and the ascent occurs along the same adiabat.

If $\gamma_v < \gamma_d$ (the usual case), both the vertical stretching (horizontal convergence; shrinking of the column: $A'/A < 1$) and ascending motion ($p'/p < 1$) increase the lapse rate, the layer thus becoming less stable. (This is the case of Figure IX-13.) Conversely, both the broadening of the layer (horizontal divergence: $A'/A > 1$) and the descending motion ($p'/p > 1$) decrease the lapse rate and therefore tend to stabilize the layer (this case is the opposite of Figure IX-13, but it can be visualized on the same figure by exchanging primed and unprimed letters).

If $\gamma_v > \gamma_d$, we would have the opposite effects, but this situation does not occur in the atmosphere, as it corresponds to absolute instability.

Equation (63) indicates that, for increasing $p'A'/pA$, the sign of the lapse rate can change, resulting in more or less pronounced inversions ($\gamma'_v < 0$). If on the contrary $p'A'/pA$ decreases tending to zero, γ'_v will tend to the dry adiabatic lapse rate γ_d.

These cases occur in large scale cyclones and anticyclones. In anticyclones, the subsidence frequently gives rise to inversions. In cyclones, the convergence aloft may lead to nearly-adiabatic lapse rates.

In the example of Figure IX-13, both effects (rising motion and shrinking) tend to make the layer more unstable, which becomes apparent by the increase in lapse rate from BC to $B'C'$.

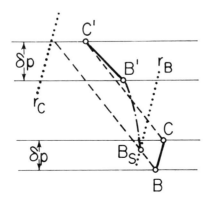

Fig. IX-14. Potentially unstable layer becoming partly saturated.

9.10.2. PART OF THE LAYER BECOMES SATURATED DURING THE ASCENT

We shall not study this case analytically, but it is easy to see the qualitative effects on a diagram.

If we assume that $\delta\theta_{aw}/\delta z < 0$, i.e., that the layer is more humid at the base than at the top, the base will become saturated (BB_s) before the top in a rising motion. From that moment it will follow the saturated adiabat (B_sB'), while the top continues along a dry adiabat (CC'). This is shown schematically on a tephigram in Figure IX-14, where it is easy to see that the ascent makes the layer absolutely unstable. This is expressed by saying that the layer was originally *potentially unstable*.

If, on the contrary, $\delta\theta_{aw}/\delta z > 0$, we have the opposite effect, as shown in Figure IX-15, and the layer is said to be *potentially stable*.

If $\delta\theta_{aw}/\delta z = 0$, the layer is said to be potentially neutral.*

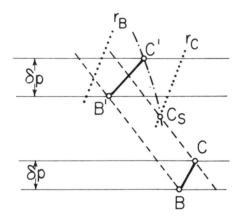

Fig. IX-15. Potentially stable layer becoming partly saturated.

In the case of bulk ascent of an atmospheric layer leading to saturation, a potentially stable layer will tend to form stratiform clouds, while a potentially unstable layer will produce cumuliform clouds and perhaps eventually convective precipitation (showers).

All the analyses of vertical stability considered so far have been summarized, for convenience, in Table IX-1.

---

* We notice that the conditions $\delta\theta_{aw}/\delta z \lessgtr 0$ refer here to the stratification of a layer, initially not saturated, which rises in the atmosphere, while conditions (44) referred to the environment, when a saturated parcel rises in it.

TABLE IX-1

Vertical stability

(A) Local stability conditions

Parcel method: $\gamma_v \lessgtr \gamma'_v = \begin{cases} \gamma_d, \text{ unsaturated}; & \text{or } \dfrac{\delta\theta}{\delta\phi} \lessgtr 0 \\ \gamma_w, \text{ saturated}; & \text{or } \dfrac{\delta\theta_{aw}}{\delta\phi} \lessgtr 0 \end{cases}$

Layer method: $(\gamma_v - \gamma') \gtrless \dfrac{A'}{A}(\gamma - \gamma_v)$

where upper inequality signs correspond to instability, lower inequality signs correspond to stability, equality signs correspond to indifference.

$\gamma_v > \gamma_d$ absolute instability
$\gamma_d > \gamma_v > \gamma_w$ conditional instability → (B)
$\gamma_v < \gamma_w$ absolute stability

(B) Finite vertical displacements: *latent instability*

$A_+ > A_-$ real latent type
$A_+ < A_-$ pseudolatent type
$A_+ = 0$ stable type

(C) Layer vertical displacements: *potential or convective instability*.

Unsaturated: $\gamma'_v = \gamma_v + (\gamma_d - \gamma_v)\left(1 - \dfrac{p'A'}{pA}\right)$

$\left.\begin{array}{l} p'/p < 0 \quad \text{ascent} \\ A'/A < 0 \quad \text{lateral convergence} \end{array}\right\} \to \text{destabilization}$

$\left.\begin{array}{l} p'/p > 0 \quad \text{descent} \\ A'/A > 0 \quad \text{lateral divergence} \end{array}\right\} \to \text{stabilization}$

With saturation: $\dfrac{\delta\theta_{aw}}{\delta\phi} \lessgtr 0 \quad \begin{cases} \text{potentially unstable (upper sign)} \\ \text{potentially stable (lower sign)} \end{cases}$

9.11. Processes Producing Stability Changes for Dry Air

From Equation (40), (Chapter VIII, Equation (21)) and the gas law, it follows that, for dry air,

$$\frac{\delta \ln \theta}{\delta p} = -\frac{R_d}{p}(\gamma_d - \gamma). \tag{64}$$

This equation suggests that a useful stability parameter could be defined by the relation

$$\sigma = -\frac{\delta \ln \theta}{\delta p}, \tag{65}$$

with the associated stability criteria

$$\sigma \gtrless 0, \tag{66}$$

where, now, the sign $>$ corresponds to stability and $<$ to instability. At a fixed pressure, σ is a single-valued function of γ. Moreover, the finite-difference approximation to σ, $-\Delta \ln\theta/\Delta p$, is obtainable immediately from a tephigram curve.

At a fixed isobaric level, σ increases as the lapse rate decreases, i.e., as the stability increases; and σ decreases as the lapse rate increases, i.e., as the stability decreases. Thus, for increasing stability $(\partial\sigma/\partial t)_p > 0$, for decreasing stability, $(\partial\sigma/\partial t)_p < 0$. From Equation (65),

$$\left(\frac{\partial\sigma}{\partial t}\right)_p = \left[\frac{\partial}{\partial t}\left(-\frac{\partial \ln\theta}{\partial p}\right)\right]_p = -\frac{\partial}{\partial p}\left(\frac{\partial \ln\theta}{\partial t}\right)_p. \tag{67}$$

Our basic coordinate set is (x, y, p, t) and in terms of this set:

$$\frac{d \ln\theta}{dt} = \left(\frac{\partial \ln\theta}{\partial t}\right)_p + u\left(\frac{\partial \ln\theta}{\partial x}\right)_p + v\left(\frac{\partial \ln\theta}{\partial y}\right)_p + \frac{dp}{dt}\frac{\partial \ln\theta}{\partial p} \tag{68}$$

where u and v represent the x- and y-components of wind velocity (positive to east and north, respectively) and dp/dt represents the vertical motion in isobaric coordinates, the total rate of change of pressure with time for a moving element of air. It can be related to the vertical velocity by an expansion of the total derivative in terms of the (x, y, z, t) coordinate set:

$$\frac{dp}{dt} = \frac{\partial p}{\partial t} + u\frac{\partial p}{\partial x} + v\frac{\partial p}{\partial y} + w\frac{\partial p}{\partial z} = \left(\frac{\partial p}{\partial t} + u\frac{\partial p}{\partial x} + v\frac{\partial p}{\partial y}\right) - g\varrho w \cong -g\varrho w. \tag{69}$$

The bracketed terms are relatively small and physically insignificant, as far as approach to saturation is concerned, i.e., they result in very small adiabatic temperature changes.

For dry adiabatic changes, $d \ln\theta/dt = 0$, and in general, from the first principle of thermodynamics,

$$\frac{d \ln\theta}{dt} = \frac{1}{c_p T}\frac{\delta Q}{\delta t}. \tag{70}$$

From Equations (65), (68) and (70),

$$\left(\frac{\partial \ln\theta}{\partial t}\right)_p = -\left[u\left(\frac{\partial \ln\theta}{\partial x}\right)_p + v\left(\frac{\partial \ln\theta}{\partial y}\right)_p\right] + \sigma\frac{dp}{dt} + \frac{1}{c_p T}\frac{\delta Q}{\delta t}. \tag{71}$$

From Equations (67) and (71),

$$\left(\frac{\partial\sigma}{\partial t}\right)_p = \frac{\partial}{\partial p}\left[u\left(\frac{\partial \ln\theta}{\partial x}\right)_p + v\left(\frac{\partial \ln\theta}{\partial y}\right)_p\right] - \frac{\partial}{\partial p}\left(\sigma\frac{dp}{dt}\right) - \frac{\partial}{\partial p}\left(\frac{1}{c_p T}\frac{\delta Q}{\delta t}\right). \tag{72}$$

The first term can be interpreted in two alternative ways. Since

$$d \ln \theta = d \ln T - \varkappa \, d \ln p, \tag{73}$$

the first term can also be written as

$$\frac{\partial}{\partial p}\left[u\left(\frac{\partial \ln T}{\partial x}\right)_p + v\left(\frac{\partial \ln T}{\partial y}\right)_p\right],$$

and is thus related to the variation in the vertical of the horizontal (strictly speaking, isobaric) temperature advection. Thus this term could be called the differential isobaric advection term. It can be interpreted in a slightly different way by expansion, i.e.

$$\frac{\partial}{\partial p}\left[u\left(\frac{\partial \ln \theta}{\partial x}\right)_p + v\left(\frac{\partial \ln \theta}{\partial y}\right)_p\right] = \left[\frac{\partial u}{\partial p}\left(\frac{\partial \ln \theta}{\partial x}\right)_p + \frac{\partial v}{\partial p}\left(\frac{\partial \ln \theta}{\partial y}\right)_p\right] +$$

$$+ \left[u\frac{\partial}{\partial p}\left(\frac{\partial \ln \theta}{\partial x}\right)_p + v\frac{\partial}{\partial p}\left(\frac{\partial \ln \theta}{\partial y}\right)_p\right]. \tag{74}$$

The second bracketed quantity can be rewritten, by interchanging the order of differentiation, as

$$u\left[\frac{\partial}{\partial x}\left(\frac{\partial \ln \theta}{\partial p}\right)\right]_p + v\left[\frac{\partial}{\partial y}\left(\frac{\partial \ln \theta}{\partial p}\right)\right]_p = -\left[u\left(\frac{\partial \sigma}{\partial x}\right)_p + v\left(\frac{\partial \sigma}{\partial y}\right)_p\right] \tag{75}$$

which is simply the isobaric advection of stability, a quasi-horizontal process. In the first bracketed quantity, $\ln \theta$ can again be replaced by $\ln T$ and the thermal wind equations of dynamic meteorology employed to introduce the isobaric shear of the geostrophic wind components – $\partial u_g/\partial p$ and $\partial v_g/\partial p$, i.e.,

$$\frac{\partial u}{\partial p}\left(\frac{\partial \ln T}{\partial x}\right)_p + \frac{\partial v}{\partial p}\left(\frac{\partial \ln T}{\partial y}\right)_p = -f\varrho\left(\frac{\partial u}{\partial p}\frac{\partial v_g}{\partial p} - \frac{\partial v}{\partial p}\frac{\partial u_g}{\partial p}\right), \tag{76}$$

where we denote by f the Coriolis parameter, $2\omega \sin\varphi$. This expression has a non-zero value only if there are cross-contour (i.e., acceleration) components of the horizontal wind field, and such effects are generally of minor importance. Thus, for quasi-geostrophic or quasi-gradient winds, the differential isobaric advection term implies merely the isobaric advection of stability.

For the benefit of those unfamiliar with the concept of the geostrophic wind, we will digress at this point to present a few formulae which illustrate and explain this concept, and from which the relation of the temperature field to the variation in the vertical of the geostrophic wind can be deduced. If we derive the equations of motion for a set of axes anchored to a rotating earth, neglect friction but take into account forces resulting from horizontal variations of pressure, we obtain the following wind-component equations:

$$u = u_g - \frac{1}{f}\frac{dv}{dt}; \qquad v = v_g + \frac{1}{f}\frac{du}{dt}. \tag{77}$$

Here u and v are the x- and y-components of the (actual) horizontal wind, and u_g and v_g the corresponding components of the geostrophic wind, a synthetic or theoretical wind, whose calculation is outlined below, and which serves as a very useful and simple approximation to the true wind. f denotes the Coriolis parameter, $2\omega \sin \varphi$ (ω = angular velocity of the earth, φ = latitude).

From set (77) we see that if there are no accelerations for either component (steady motion), the wind and the geostrophic wind, are identical. The latter can be obtained from the following equations, in which the hydrostatic equation has been used to convert horizontal gradients of pressure to isobaric gradients of geopotential:

$$u_g = -\frac{1}{f\varrho}\frac{\partial p}{\partial y} = -\frac{1}{f}\left(\frac{\partial \varphi}{\partial y}\right)_p; \qquad v_g = \frac{1}{f\varrho}\frac{\partial p}{\partial x} = \frac{1}{f}\left(\frac{\partial \varphi}{\partial x}\right)_p. \tag{78}$$

Since f at mid-latitudes is of the order of 10^{-4} s^{-1}, it follows from Equation (77) that the difference between the actual wind and the geostrophic wind, in any direction, has a magnitude closely comparable to that of the change of the actual wind component in any orthogonal direction, following the motion, for a period of 3 h. Since a reasonable value of acceleration could be 50% of the velocity per day, departures of winds from geostrophic values would average less than 10%. For balanced motion in a circular path (total acceleration zero), the wind is known as the gradient wind.

The isobaric shear of the geostrophic wind (which is also a first approximation to the isobaric shear of the actual wind) can be found from set (78), by a differentiation with respect to pressure, making use of the hydrostatic equation and the ideal gas law:

$$\frac{\partial u_g}{\partial p} = -\frac{1}{f}\frac{\partial}{\partial p}\left(\frac{\partial \varphi}{\partial y}\right)_p = -\frac{1}{f}\left[\frac{\partial}{\partial y}\left(\frac{\partial \varphi}{\partial p}\right)\right]_p = \frac{1}{f}\left[\frac{\partial(1/\varrho)}{\partial y}\right]_p = \frac{R_d}{fp}\left(\frac{\partial T}{\partial y}\right)_p \tag{79}$$

$$\frac{\partial v_g}{\partial p} = \frac{1}{f}\frac{\partial}{\partial p}\left(\frac{\partial \varphi}{\partial x}\right)_p = \frac{1}{f}\left[\frac{\partial}{\partial x}\left(\frac{\partial \varphi}{\partial p}\right)\right]_p = -\frac{1}{f}\left[\frac{\partial(1/\varrho)}{\partial x}\right]_p = -\frac{R_d}{fp}\left(\frac{\partial T}{\partial x}\right)_p. \tag{80}$$

Equations (79) and (80) are known as the thermal wind equations, since they relate the thermal field to the change of geostrophic wind in the vertical. These are the equations employed to produce the result quoted in Equation (76).

The second term in the stability-parameter tendency Equation (72), the vertical motion term, can best be studied after further expansion into the form:

$$-\frac{\partial}{\partial p}\left(\sigma \frac{dp}{dt}\right) = -\frac{dp}{dt}\frac{\partial \sigma}{\partial p} - \sigma \frac{\partial}{\partial p}\left(\frac{dp}{dt}\right). \tag{81}$$

The first of these two terms represents simply the vertical advection of stability, while the second represents the effects of the vertical shrinking or stretching of an air column, i.e., the change in $\sigma = -\Delta \ln\theta/\Delta p$ when $\Delta \ln\theta$ remains constant, for dry adiabatic changes, but Δp changes (see Section 10.1.).

The third term in the stability tendency equation expresses the effect of vertical variations in non-adiabatic heating. In the free atmosphere, away from clouds or the Earth's surface, this effect will be small, except possibly near the tropopause or similar discontinuity in lapse rate and hence in the degree of turbulent mixing. Near the ground, the radiative $\delta Q/\delta t$ decreases in magnitude with height, as was pointed out in an earlier section, so that a pronounced diurnal cycle of the lapse rate exists in the lower layers of the atmosphere.

Before considering illustrations on schematic tephigrams of the vertical motion processes, it will be instructive to examine the degree of conservatism of stability, as measured by the parameter σ, and to consider typical vertical-motion distributions in the vertical. We shall first obtain a relation for the change of stability following the motion, $d\sigma/dt$, for quasi-gradient flow and quasi-adiabatic conditions. Collecting component terms for $(\partial \sigma/\partial t)_p$ from the expansions in Equations (72), (74), (75), (76), and (81), it follows that

$$\frac{d\sigma}{dt} = \left(\frac{\partial \sigma}{\partial t}\right)_p + u\left(\frac{\partial \sigma}{\partial x}\right)_p + v\left(\frac{\partial \sigma}{\partial y}\right)_p + \frac{dp}{dt}\frac{\partial \sigma}{\partial p} = -\sigma \frac{\partial}{\partial p}\left(\frac{dp}{dt}\right). \tag{82}$$

From the hydrodynamical equation of continuity (a mathematical expression of the law of conservation of mass), the shrinking or stretching of a moving air column can be related to the accumulation or depletion of mass associated with velocity gradients in isobaric surfaces such that

$$-\frac{\partial}{\partial p}\left(\frac{dp}{dt}\right) = \left(\frac{\partial u}{\partial x}\right)_p + \left(\frac{\partial v}{\partial y}\right)_p. \tag{83}$$

Equation (83), the equation of continuity in isobaric coordinates, expresses the fact that there can be no mass change between fixed isobaric surfaces. The quantity on the right is known as the isobaric divergence, when positive, and the isobaric convergence, when negative, these terms implying depletion and accumulation of mass, respectively. The expression is also referred to, quite generally, as the isobaric velocity divergence, and expresses the rate of change of mass by quasi-horizontal motions (outflow minus inflow). The term on the left of Equation (83) expresses the effect of motions through isobaric surfaces in changing the mass of a layer.

It follows from Equations (82) and (83) that

$$\frac{d \ln \sigma}{dt} = \left(\frac{\partial u}{\partial x}\right)_p + \left(\frac{\partial v}{\partial y}\right)_p. \tag{84}$$

Thus isobaric divergence corresponds to an increase in magnitude of σ, and isobaric convergence to a decrease in magnitude of σ. It is clear, however, that no mechanism exists for changing stability to instability in a moving element of air, and that instability can only be produced locally if it existed previously upstream. Since instability will rapidly be destroyed by convective turbulent mixing (leading to $\gamma = \gamma_d$), it follows that super-adiabatic lapse rates must have an extremely ephemeral existence. However, it should be emphasized that significant consequences of the release of instability are generally associated only with the condensation process (convective clouds, showers and thunderstorms). For saturated air there are different criteria for instability – the release of latent heat energy must be taken into account – and in general a combined dynamic–thermodynamic analysis is required. This topic will be considered further in the next section.

From Equation (69), $dp/dt \cong -g\varrho w$, it follows that, over level ground, dp/dt is essentially zero at the surface. Integrating the equation of continuity from the surface (p_0) to an arbitrary level:

$$\frac{dp}{dt} = \int_p^{p_0} \left(\frac{\partial u}{\partial x} + \frac{\partial v}{\partial y}\right)_p dp. \tag{85}$$

Thus, in a stratum of divergence at and above the surface, dp/dt will be positive, corresponding to subsidence ($w<0$); in a stratum of convergence at and above the surface, dp/dt will be negative, corresponding to ascent ($w>0$). Vertical motion studies have revealed that on the average a rather simple pattern of vertical motions may be expected in the troposphere. In general, tropospheric vertical motions tend to have the same sign, with a maximum absolute value in the mid-troposphere, in the vicinity of the 600 mb level, on the average. At this level of maximum (or minimum) dp/dt, the isobaric divergence must be zero, and this level is generally referred to as the level of nondivergence (L.N.D.). In any individual case, this level may depart appreciably from 600 mb, and does not in any event coincide with a specific isobaric surface, over any great area, at least. Moreover, especially in complex atmospheric situations, there may well be more than one level of non-divergence in the troposphere. However, the broad behavior of the troposphere is usually consistent with the existence of a single level of non-divergence in the mid-troposphere, either with divergence below and convergence above (subsidence) or with convergence below and divergence above (ascent).

We have already seen that divergence in a stable atmosphere will tend to increase the stability (decrease the lapse rate) and that convergence in a stable atmosphere will tend to decrease the stability (increase the lapse rate), both such rates being proportional to the initial value of the stability. Thus, for small stability (($\gamma_d - \gamma$) small) the effects of a given divergence or convergence on the stability will be small, while for large stability (($\gamma_d - \gamma$) large) the effects will be much greater (cf. Section 10.1). It is of interest to apply this concept, together with the normal divergence-convergence pattern of the atmosphere, to a consideration of the relative sharpness of frontal discontinuities as seen on atmospheric soundings on aerological diagrams. The significant discontinuity, charted as the front on quasi-horizontal analyses, is the upper boundary of the transition zone of intense isobaric thermal gradients, separating the stratum of great stability (σ large) characterizing this zone from the warm air mass above of normal stability (σ relatively small). The discontinuity in stability at this level may be taken as a measure of the sharpness of the front on the plotted sounding.

Since isobaric divergence increases the stability in the transition zone more rapidly than in the warm air, it will sharpen the frontal discontinuity. Since isobaric convergence decreases the stability in the transition zone more rapidly than in the warm air, it will weaken the frontal discontinuity. It is reasonable to expect that these effects would be visible, at least on the average, and synoptic experience does indeed verify the following conclusions.

Tropospheric ascent:

$$\left\{\begin{array}{l}\text{cold front, slow moving with persistent weather}\\ \text{warm front, normal activity re weather}\end{array}\right\}\left\{\begin{array}{l}\text{convergence below L.N.D.}\\ \text{divergence above L.N.D.}\end{array}\right\}$$

$$\left\{\begin{array}{l}\text{lower troposphere – frontal discontinuity relatively weak}\\ \text{upper troposphere – frontal discontinuity relatively sharp}\end{array}\right\}.$$

Tropospheric subsidence:

$$\left\{\begin{array}{l}\text{cold front, rapidly moving with rapid clearing}\\ \text{warm front, virtually no weather}\end{array}\right\}\left\{\begin{array}{l}\text{divergence below L.N.D.}\\ \text{convergence above L.N.D.}\end{array}\right\}$$

$$\left\{\begin{array}{l}\text{lower troposphere – frontal discontinuity relatively sharp}\\ \text{upper troposphere – frontal discontinuity relatively weak}\end{array}\right\}.$$

Let us consider now a typical situation with ascent throughout the troposphere, with isobaric convergence in the lower troposphere and divergence in the upper troposphere, as it might appear on an aerological diagram, neglecting horizontal advective processes (Figure IX-16).

In the upper troposphere $\partial/\partial p(\mathrm{d}p/\mathrm{d}t)<0$, and $\sigma>0$, so $\partial\sigma/\partial t>0$ (increasing stability). In the lower troposphere $\partial/\partial p(\mathrm{d}p/\mathrm{d}t)>0$, and $\sigma>0$, so $\partial\sigma/\partial t<0$ (decreasing stability).

Two weak discontinuities in lapse rate (and stability) are produced by this pattern of ascending motion and adiabatic cooling, and the new sounding may appear to contain a frontal discontinuity (at high levels), comparable to the passage of a weak

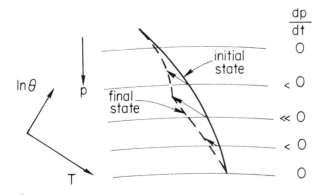

Fig. IX-16. Stability changes with isobaric convergence in the lower troposphere and divergence in the upper troposphere.

cold front. This phenomenon is often observed in advance of cold fronts especially when the warm air mass is relatively dry so that dry adiabatic cooling can take place; for this reason the phenomenon is known as 'pre-frontal cooling'. It can be distinguished from true frontal cooling, with a new and colder air mass below the front, by two distinguishing characteristics. In the first place, the cooling is relatively small in the lower portion of the sounding (effectively zero at the ground in the absence of advection), and, in the second place, the humidity increases at any level as the temperature drops. Thus it is always possible to obtain the new sounding by the ascent of typical warm air.

Let us consider now a typical subsidence situation, illustrated on a tephigram (Figure IX-17).

In the upper troposphere, $\partial/\partial p(\mathrm{d}p/\mathrm{d}t) > 0$ and $\sigma > 0$, so that $\partial\sigma/\partial t < 0$ (decreasing stability). In the lower troposphere, $\partial/\partial p(\mathrm{d}p/\mathrm{d}t) < 0$ and $\sigma > 0$, so that $\partial\sigma/\partial t > 0$ (increasing stability). Once again a discontinuity in lapse rate has been produced by the pattern of descending motion and adiabatic warming, somewhat comparable to a warm front situation. Since the inversion has been produced by subsidence, it is known as a subsidence inversion. It can be distinguished from a true warm frontal

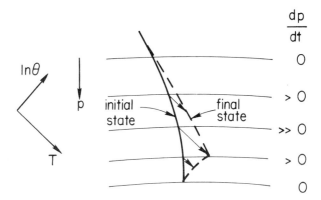

Fig. IX-17. Stability changes associated with subsidence.

passage by the relatively slight warming in the upper troposphere, the steep lapse rate (small stability) above the inversion, and by the low humidities that will appear near the top of the inversion layer. One can, of course, obtain representative temperatures and dewpoints by the ascent of the relatively dry air (in nature, or by carrying it out on the diagram).

9.12. Stability Parameters of Saturated and Unsaturated Air, and Their Time Changes

If we had maintained the effects of water vapor in the derivations of the last section (i.e., starting from Equation (40)), we might have defined a virtual stability parameter for unsaturated air as

$$\sigma_v = -\frac{\delta \ln \theta_v}{\delta p} = \frac{R_d}{p}(\gamma_d - \gamma_v). \tag{86}$$

All the relations in Section 11 would still be valid, with the subscript v as above, with quite acceptable accuracy. For most practical purposes, such refinements are unnecessary since the virtual temperature increments for a vertically-moving parcel and for its environment at any time would differ by only a few tenths of a degree Celsius, in general.

When dealing with saturated air, numerous complications arise, which generally introduce uncertainties in the temperature (or virtual temperature) of a parcel which exceed the virtual temperature increment differential between parcel and environment. In the first place, one will not know whether ascent is of the saturated adiabatic or pseudo-adiabatic category. Initial saturation is almost invariably followed by cloud development without precipitation (a saturated adiabatic process), and only later, when additional physical criteria are satisfied, may precipitation occur (a pseudo-adiabatic process). For the truly reversible process, the (p, T) variation depends on the initial condensation level and the virtual temperature will now depend on the concentration of condensed water as well as of water vapor. Since the volume occupied by the condensed phase is negligible, the virtual temperature of the cloud air (incorporating the effect of the condensed phase as a component of the air) is given by an extension of the treatment in Chapter IV, Section 11 as

$$T_v = \frac{T(1 + r_w/\varepsilon)}{(1 + r_w)} \frac{1 + r_w}{1 + r_w + r_l} = \frac{T(1 + r_w/\varepsilon)}{(1 + r_t)}. \tag{87}$$

These various effects are by no means compensatory. Moreover, once the temperature of an ascending parcel falls below $0\,^\circ\text{C}$, additional uncertainties are introduced with respect to the level (if any) at which the transition from water-saturation to ice-saturation takes place. If we imagine a large ascending mass of air, of cloud dimensions, the micro-processes that take place inside the cloud will greatly affect stability considerations. We must conclude, then, that the stability or instability of saturated air can only be handled in its grosser aspects by a pure thermodynamic analysis.

If we have a parcel of saturated air, it will be stable relative to its environment (saturated or unsaturated) provided that $\gamma < \gamma_s$, where γ_s is some sort of saturated adiabatic lapse rate (true or pseudo, relative to water or ice). If $\gamma > \gamma_s$, the parcel will exhibit instability (analogous to dry-air instability when $\gamma > \gamma_d$). Since $\gamma_s < \gamma_d$, a layer of unsaturated air may itself be stable, but will still accelerate the vertical motions of a saturated parcel, if $\gamma_s < \gamma < \gamma_d$. This we have defined (in Section 5) as conditional instability, since it depends on the state of a moving parcel. If it is possible for a moving air parcel, on adiabatic ascent to and beyond saturation, to achieve the temperature of the environment at some level, the instability is then known as latent instability (see Section 6). If the conditional instability criterion $(\gamma_d > \gamma > \gamma_s)$ is to be expressed in terms of the vertical gradient of some specific property of the air, the appropriate property must relate to the saturation or pseudo-adiabat through (p, T).

These adiabats are labelled in terms of θ_w, (we may drop the subscript 'a' here without ambiguity), but the θ_w pseudo-adiabat only passes through (p, T) if the air is saturated. Let us define the saturation potential temperature, θ_s, as the value of θ_w for the pseudo-adiabat passing through (p, T), i.e., as the value of θ_w if the air were saturated. Mathematically,

$$\theta_s(p, T, r) = \theta_w(p, T, r_w), \quad \text{for} \quad r_w = r_w(p, T). \tag{88}$$

This is a rather artificial property, since it cannot be achieved by any simple physical process. It is, nevertheless, a rather useful concept, especially in air-mass analysis. Since most air masses have lapse rates close to the pseudo-adiabatic, a given frontal surface will be characterized by nearly constant values of θ_s. θ_s is conservative for saturated adiabatic processes, since it is then equal to θ_w, but is not conservative for dry adiabatic processes.

We therefore have conditional instability in any layer for which θ_s decreases with height, or for which

$$\sigma_s = -\frac{\delta \ln \theta_s}{\delta p} < 0. \tag{89}$$

We will have latent instability in any layer for which θ_s is less than θ_w at some lower level, as is evident from Figure IX-5 or any sounding on a thermodynamic diagram. Conditional instability without latent instability is essentially no instability at all; on the other hand, latent instability is impossible without conditional instability. Thus the presence of conditional instability indicates that one should examine the sounding more carefully for the possible existence of latent instability; if conditional instability is absent or marginal, a further stability analysis is not required.

One could carry out an analysis of stability changes with time in terms of the parameter, σ_s, so that $\partial \sigma_s / \partial t$ would correspond to decreasing stability or increasing conditional instability. However, since the only important class of conditional instability is latent instability, time changes of this latter phenomenon are more revealing and significant. Since the θ_s values and the θ_w values involved in latent instability refer to quite variable layers, in practice we would want to investigate $\partial \theta_s / \partial t$ in the middle troposphere and $\partial \theta_w / \partial t$ in the lower troposphere. Pressure levels of 500 and 850 mb, respectively, have been adopted for the Showalter Stability Index, useful for hail, tornado and thunderstorm forecasting; this index was defined at the end of Section 6. The Showalter Index is comparable to, but not identical with: $(\theta_s)_{500} - (\theta_w)_{850}$. It will always have the same sign as the above quantity; negative values are usually associated with severe thunderstorm situations.

If we have a layer of saturated air, there will be absolute instability if $\gamma > \gamma_s$ or if

$$\sigma_w = -\frac{\delta \ln \theta_w}{\delta p} < 0. \tag{90}$$

This instability would soon be released by turbulent overturning and mixing of the cloud air, a process that occurs in convective-type clouds, in which individual cells are often characterized by marked ascent or descent. It follows that saturated air is seldom observed with a lapse rate exceeding γ_s; but $\partial\sigma_w/\partial t < 0$ in saturated air would, of course, indicate the development of convective clouds, which are often observed to be imbedded in stratiform cloud decks.

Now let us consider the case of a layer of unsaturated air for which $\sigma_w < 0$, but $\sigma > 0$ and σ_s may be positive or negative. We may note in passing that if $\sigma_w > 0$ at all levels, θ_w can never exceed θ_s at a higher level, even though layers of conditional instability exist, since $\theta_s \geqslant \theta_w$. Thus, layers with $\sigma_w < 0$ and with $\sigma_s < 0$ (not necessarily the same layers) are necessary but not sufficient conditions for latent instability. If the layer with $\sigma_w < 0$ is now lifted to saturation, by a general ascent over a large area, instability can be released as soon as saturation is achieved. This we have defined in Section 10 as potential instability and it can only be realized by mass ascent. Latent instability, on the other hand, can be realized by parcel ascent, usually a convective or thermal process. For large scale energy release, however, general ascending motion is usually required; in subsiding air, convective activity is usually choked off and convective clouds seldom penetrate a subsidence inversion, for example.

In order to investigate $\partial\theta_s/\partial t$, $\partial\sigma_s/\partial t$, $\partial\theta_w/\partial t$ or $\partial\sigma_w/\partial t$, we require an analytic formulation for $\theta_w(p, T, r)$ and of $\theta_w(p, T, r_w)$, and thus must be able to integrate the pseudo-adiabatic equation, at least approximately, from p to $p_0 = 1000$ mb, and solve, directly or indirectly, for θ_w. Since we will be interested basically in $\partial\theta_w/\partial t$ and $\partial\theta_w/\partial p$, we may introduce approximations that would not be permitted if precise values of θ_w, alone, were to be the end product.

The simplest way to formulate θ_w or θ_s is to make use of the approximate equality between isobaric and adiabatic wet-bulb temperatures. We will assume that $T_{aw} = T_{iw} = T_w$ and $\theta_{aw} = \theta_{iw} = \theta_w$, where θ_w is the isobaric wet-bulb temperature of moist air at (p_0, θ, r) – for purposes of calculation – but is also considered to lie on the

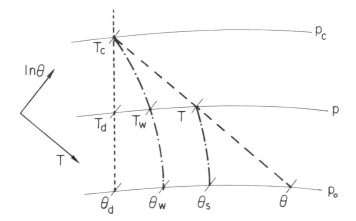

Fig. IX-18. Temperature parameters, on a tephigram.

same pseudo-adiabat as (p_c, T_c) and (p, T_{aw}), as illustrated on Figure IX-18, where dash-dot lines represent pseudo-adiabats and the dotted line a saturation mixing ratio line, corresponding to the actual mixing ratio of the air sample at (p, T). Along it, we can assert

$$r = r_w(p_c, T_c) = r_w(p, T_d) = r_w(p_0, \theta_d), \tag{91}$$

where we have defined a parameter θ_d, the dew-point potential temperature.

Consistent with the assumed equality of isobaric and adiabatic wet-bulb temperatures and potential temperatures, we may employ a simplified version of the Psychrometric Equation (Chapter VII, Equation (30)) and write, at p_0,

$$e_w(\theta_d) = e_w(\theta_w) - \frac{c_{pd} p_0}{\varepsilon l_v} (\theta - \theta_w). \tag{92}$$

We may regard $\varepsilon l_v / c_{pd} p_0$ as a constant and denote it by B, i.e.

$$B = \frac{\varepsilon l_v}{c_{pd} p_0}. \tag{93}$$

Since

$$\frac{r}{r + \varepsilon} = \frac{e_w(\theta_d)}{p_0} = \frac{e_w(T_d)}{p} \tag{94}$$

we have, from Equations (92), (93) and (94),

$$\theta_w = \theta - B\left[e_w(\theta_w) - \frac{p_0}{p} e_w(T_d)\right]. \tag{95}$$

If the air at (p, T) is assumed to be saturated, as implied by the definition of θ_s, we have, by analogy,

$$\theta_s = \theta - B\left[e_w(\theta_s) - \frac{p_0}{p} e_w(T)\right]. \tag{96}$$

Separating out the properties of the original sample of moist air $-p$, T, T_d and θ, we can rewrite Equations (95) and (96) as

$$\theta_w + B e_w(\theta_w) = \theta + B \frac{p_0}{p} e_w(T_d) = \theta_e \tag{97}$$

$$\theta_s + B e_w(\theta_s) = \theta + B \frac{p_0}{p} e_w(T) = \theta_{se}. \tag{98}$$

In Equation (97), θ_e, a single-valued function of θ_w, is *not* the isobaric equivalent potential temperature, θ_{ie}, defined in (Chapter VII, Section 14), but is instead the isobaric equivalent temperature, at $p_0 = 1000$ mb, of air which has been taken dry-adiabatically to 1000 mb. In other words, θ_{ie} is not strictly conservative for dry adiabatic changes of state, whereas θ_{ae} is conservative for such a process (and so also is θ_e as defined by Equation (97)). These facts become apparent if we introduce Equation (93) into (Chapter VII, Equation (29)), giving

$$T_{ie} = T + B\frac{p_0}{p}e = T + B\frac{p_0}{p}e_w(T_d). \tag{99}$$

Introducing the definition of potential temperature, we have

$$\theta_{ie} = T_{ie}\left(\frac{p_0}{p}\right)^\varkappa = \theta + B\left(\frac{p_0}{p}\right)^{1+\varkappa}e_w(T_d) \tag{100}$$

whereas Equation (99) applied at p_0 gives, using Equation (94),

$$\theta_e = \theta + Be_w(\theta_d) = \theta + B\frac{p_0}{p}e_w(T_d),$$

which is identical to Equation (97). θ_e is much more useful than θ_{ie}, being conservative for a dry adiabatic process (for which θ and r are, in turn, conserved), and could be termed the isobaric-adiabatic equivalent potential temperature. There will be small changes in θ_e (but not in θ_{ae}) for saturated adiabatic ascent or descent, but these are less important from a practical standpoint in view of the uncertainties in the precise physical processes in operation.

In Equation (98), θ_{se}, a single-valued function of θ_s, may be called the saturation equivalent potential temperature. It depends only on the pressure and temperature of an air sample and can be computed, simply and directly in an analytic sense, from those parameters. Both θ_e and θ_{se} are ideally suited for a computer analysis of stability characteristics.

Now we may return to a consideration of stability criteria and stability changes, for the three most interesting classes of instability – conditional, potential and latent. Analogous to Equations (89) and (90), we may define

$$\sigma_e = -\frac{\delta \ln \theta_e}{\delta p} \quad \text{and} \quad \sigma_{se} = -\frac{\delta \ln \theta_{se}}{\delta p}. \tag{101}$$

We have conditional instability if

$$\gamma > \gamma_s \quad \text{or} \quad \sigma_s < 0 \quad \text{or} \quad \sigma_{se} < 0.$$

We have potential instability if

$$\sigma_w < 0 \quad \text{or} \quad \sigma_e < 0.$$

We have latent instability if

$$\theta_w > \theta_s(p') \quad \text{for some} \quad p' < p,$$

or if

$$\theta_e > \theta_{se}(p') \quad \text{for some} \quad p' < p.$$

We may now consider the physical processes that cause changes in the various types of instability being considered in this section. It will be convenient in formulating relations for such changes to neglect non-adiabatic effects, so that such effects will be discussed briefly first, in qualitative terms. Turbulent diffusion of heat and of water vapor in the vertical will always act to reduce potential stability if initially there is stability (and to reduce potential instability if initially there is instability), since the net result of vertical mixing processes is to reduce the vertical gradient of θ_e. On the other hand, turbulent interchange will favor the development of conditional instability. The other major non-adiabatic process is radiation, chiefly the long wave radiation of the earth and its atmosphere since solar radiation is absorbed primarily at the earth's surface, promoting instability of all types in the lower layers of the atmosphere. To appreciate the effects of long-wave (infrared) radiative exchange, we require information on the isobaric temperature changes which result from the vertical divergence of infrared radiative fluxes. When skies are clear, one can expect a cooling of the order of 1 to 2 °C day^{-1} in the mid- and upper troposphere, somewhat less cooling in the lower troposphere and lower stratosphere and usually a minimum of cooling at the tropopause and a maximum (second maximum) at the earth's surface. When an overcast cloud layer is present, the atmospheric cooling is enhanced immediately above the cloud and greatly reduced below the cloud (and throughout the entire atmosphere below the cloud, unless the cloud is very high). The cloud itself will cool markedly if a low cloud and negligibly if a high cloud; the upper layers of the cloud always cool while the lower layers tend to warm, the net effect almost invariably being a cooling. Thus, with clear skies the effect of radiative cooling is to stabilize the atmosphere near the ground (offset by convective and conductive processes during the day), destabilize the middle and lower troposphere, stabilize the upper troposphere and destabilize the lower stratosphere. Radiative processes, like turbulent mixing processes, tend to reduce lapse-rate discontinuities and to smooth out bases and tops of inversion layers. When clouds are present, these effects are modified. Below an overcast layer, radiation acts to stabilize the atmosphere, and similarly above the cloud layer. Within the cloud layer, radiation acts to destabilize the stratification, and this effect may often be important in the development of nocturnal thunderstorms and of convective-type middle clouds. The cloud destabilization is intensified by the strong atmospheric cooling above the cloud and the warming below the cloud.

Returning now to adiabatic processes, the development of conditional instability is essentially associated with an increasing lapse rate (and a decrease in σ). Thus the processes discussed under the heading of stability changes for dry air (Section 11) apply here with equal force, and further elaboration is unnecessary. For the development of latent instability, one requires an increase in θ_w or θ_e at some level or a decrease in T or θ_s or θ_{se} at some higher level. At the higher level, ascent and advective cooling are the factors which assist in this process, whereas at the lower level horizontal advection of air of higher θ_w or θ_e is the important factor, vertical motions merely shifting in the vertical the level of maximum θ_w (or θ_e) along the trajectory. Expressed in another way, since θ_e is conserved for all adiabatic processes, local changes are due to horizontal and vertical advection.

The rate of change of potential stability (or instability) can be investigated by formulating $\partial \sigma_e / \partial t$, as we did $\partial \sigma / \partial t$ for dry air in Section 11, Equation (72) et seq., again employing an (x, y, p, t) coordinate set. Thus, from Equation (101),

$$\frac{\partial \sigma_e}{\partial t} = \frac{\partial}{\partial t}\left(-\frac{\partial \ln \theta_e}{\partial p}\right) = -\frac{\partial}{\partial p}\frac{\partial \ln \theta_e}{\partial t}, \tag{102}$$

and $\partial \sigma_e / \partial t$ will be positive for increasing potential stability or for decreasing potential instability. Since θ_e may be considered conserved following the motion of the air,

$$\frac{d \ln \theta_e}{dt} = 0 = \frac{\partial \ln \theta_e}{\partial t} + u \frac{\partial \ln \theta_e}{\partial x} + v \frac{\partial \ln \theta_e}{\partial y} + \frac{dp}{dt}\frac{\partial \ln \theta_e}{\partial p}. \tag{103}$$

Hence, from Equations (102) and (103) and using Equation (101)

$$\frac{\partial \sigma_e}{\partial t} = \frac{\partial}{\partial p}\left(u \frac{\partial \ln \theta_e}{\partial x} + v \frac{\partial \ln \theta_e}{\partial y}\right) - \frac{\partial}{\partial p}\left(\sigma_e \frac{dp}{dt}\right). \tag{104}$$

Carrying out the differentiation implied by Equation (104), we have, again using Equation (101),

$$\frac{\partial \sigma_e}{\partial t} = \left(\frac{\partial u}{\partial p}\frac{\partial \ln \theta_e}{\partial x} + \frac{\partial v}{\partial p}\frac{\partial \ln \theta_e}{\partial y}\right) - \left(u \frac{\partial \sigma_e}{\partial x} + v \frac{\partial \sigma_e}{\partial y}\right) - \sigma_e \frac{\partial}{\partial p}\frac{dp}{dt} - \frac{dp}{dt}\frac{\partial \sigma_e}{\partial p}. \tag{105}$$

In order to evaluate the first term, above, we must differentiate Equation (97), at constant pressure. Thus,

$$\left(\frac{\partial \ln \theta_e}{\partial x}\right)_p = \frac{1}{\theta_e}\left\{\left(\frac{\partial \theta}{\partial x}\right)_p + B \frac{p_0}{p}\frac{d e_w(T_d)}{d T_d}\left(\frac{\partial T_d}{\partial x}\right)_p\right\}. \tag{106}$$

Introducing the equation for potential temperature and the Clausius-Clapeyron equation,

$$\frac{\partial \ln \theta_e}{\partial x} = \frac{1}{\theta_e}\left(\frac{p_0}{p}\right)^\kappa \frac{\partial T}{\partial x} + \frac{B p_0 \varepsilon e_w(T_d) l_v}{\theta_e p\; R_d T_d^2}\frac{\partial T_d}{\partial x}. \tag{107}$$

Writing a similar equation for $\partial \ln \theta_e / \partial y$, and using Equation (93) we have

$$\frac{\partial u}{\partial p}\frac{\partial \ln \theta_e}{\partial x} + \frac{\partial v}{\partial p}\frac{\partial \ln \theta_e}{\partial y} = \frac{1}{\theta_e}\left(\frac{p_0}{p}\right)^\kappa\left(\frac{\partial u}{\partial p}\frac{\partial T}{\partial x} + \frac{\partial v}{\partial p}\frac{\partial T}{\partial y}\right) +$$

$$+ \frac{\varepsilon^2 l_v^2 e_w(T_d)}{c_{pd} R_d \theta_e p T_d^2}\left(\frac{\partial u}{\partial p}\frac{\partial T_d}{\partial x} + \frac{\partial v}{\partial p}\frac{\partial T_d}{\partial y}\right). \tag{108}$$

The first term was examined for the dry air stability tendency analysis in Equation

(76), by introducing the isobaric shear of the geostrophic wind

$$\frac{\partial T}{\partial x} = -\frac{fp}{R_d}\frac{\partial v_g}{\partial p} \quad \text{and} \quad \frac{\partial T}{\partial y} = \frac{fp}{R_d}\frac{\partial u_g}{\partial p}. \tag{109}$$

Thus, the first term in Equation (108) vanishes for geostrophic or gradient (i.e., non-accelerated) winds, and can be neglected in general. The remaining term in Equation (108) vanishes for saturated air, since then $T_d = T$, but for unsaturated air it may be significantly positive or negative (we will return to this term shortly).

The second term in Equation (105) represents the isobaric advection of potential stability (or instability) and the fourth term the vertical advection. The remaining term can be written, using Equation (83), as

$$-\sigma_e \frac{\partial}{\partial p}\frac{dp}{dt} = \sigma_e\left(\frac{\partial u}{\partial x} + \frac{\partial v}{\partial y}\right), \tag{110}$$

and is thus the vertical shrinking or stretching term (or, the horizontal divergence or convergence term, respectively).

As in the case of dry-air stability, it is instructive to consider the degree of conservation of potential stability, following a trajectory, for the case of quasi-gradient flow and quasi-adiabatic conditions. We commence with the formal relation for $d\theta_e/dt$, i.e.,

$$\frac{d\theta_e}{dt} = \frac{\partial \theta_e}{\partial t} + u\frac{\partial \theta_e}{\partial x} + v\frac{\partial \theta_e}{\partial y} + \frac{dp}{dt}\frac{\partial \theta_e}{\partial p}. \tag{111}$$

Introducing Equation (111) into (105), substituting Equation (108) and (110), and making the geostrophic assumption via Equation (109), into wind shear terms but not convergence terms, we obtain

$$\frac{d\theta_e}{dt} = \sigma_e\left(\frac{\partial u}{\partial x} + \frac{\partial v}{\partial y}\right) + \frac{\varepsilon^2 l_v^2 e_w(T_d)}{fc_{p_d}\theta_e p^2 T_d^2}\left(\frac{\partial T}{\partial y}\frac{\partial T_d}{\partial x} - \frac{\partial T}{\partial x}\frac{\partial T_d}{\partial y}\right). \tag{112}$$

These two terms express the only mechanisms (apart from non-adiabatic effects or strongly accelerated flow) for a changing potential stability in a moving element of air. The first term predicts that divergence (by which we mean horizontal, or more strictly, isobaric divergence) will increase either potential stability or instability, whereas convergence decreases both potential stability and instability. However, such a process can never convert potential stability to potential instability, or vice versa. Thus the final term is very important, since in an air mass with potential stability everywhere it can create potential instability, under the appropriate conditions of temperature and dew point gradients (on an isobaric surface). This term can be considered to represent the effect of advection of dew point by the thermal wind (an appellation applied to the vector shear of the geostrophic wind, the orthogonal components of this shear being related to the thermal field by the set (109)).

To clarify this concept, let us recall that the (x, y) axes represent a right-handed cartesian set of axes. Conventionally they are chosen, in meteorology, with x increasing to the east and y to the north. Since the orientation is really arbitrary, let us select the x-axis as parallel to isotherms, with temperature decreasing in the direction of y increasing (see Figure IX-19, in which solid lines represent isotherms and dashed

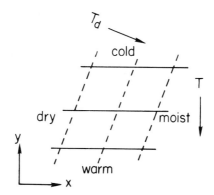

Fig. IX-19. Synoptic situation conducive to creation of potential instability.

lines represent lines of constant dew point). From Equation (109), we see that the thermal wind 'blows' in the direction of the positive x-axis (the wind component in this direction increasing with height, with the orthogonal component constant with height). From Equation (112), we see that decreasing potential stability (or increasing potential instability) requires that $\partial T_d/\partial x$ be positive, i.e., that the dew point increase in the direction of the thermal wind. In other words, the thermal wind must advect drier air, essentially with a true advection proceeding more rapidly above the given level than below. This synoptic situation is not uncommon east of the Rocky Mountains in the United States, especially in summer, with maritime tropical air from the Gulf of Mexico to the east and continental tropical air from the southwestern (arid) states to the west. When a strong thermal-wind advection of drier air takes place in a situation with general ascent to release the instability, hail and tornadoes often occur.

9.13. Radiative Processes and Their Thermodynamic Consequences

Let us consider first radiative processes near the ground, with clear skies. During most of the daylight period, the earth's surface gains energy by radiation, the solar radiation absorbed exceeding the net loss of infrared radiation. This radiative energy gain is dissipated in three ways – by heating of the ground, by heating of the air and by evaporation from the surface. Frequently, it is possible to forecast the heat input into the air, and from this energy (Q_a) one may estimate the probable maximum temperature. Let us assume that the sounding at a time of minimum temperature, $T_1(p)$, is known or can be estimated, and that the sounding at time of maximum temperature, $T_2(p)$, is characterized by a dry-adiabatic lapse rate up to the level where diurnal changes are small, as shown in Figure IX-20.

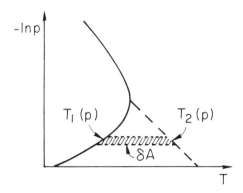

Fig. IX-20. Effect of solar radiation on temperature stratification.

Let us define an element of area, δA, on the emagram in Figure IX-20, as

$$\delta A = -(T_2 - T_1)\, d\ln p. \tag{113}$$

From the First Principle of Thermodynamics, the heat input into a cylindrical box of unit cross-section for this infinitesimally thin layer is

$$\delta Q_a = -c_p(T_2 - T_1)\frac{dp}{g}. \tag{114}$$

Introducing Equation (113) into (114) and integrating over the entire atmosphere (from surface pressure, p_s, to zero pressure)

$$Q_a = \frac{c_p}{g} \int_{p=p_s}^{p=0} p\, \delta A. \tag{115}$$

From the Theorem of the Mean, this can be rewritten as

$$Q_a = \frac{c_p}{g} \bar{p} A. \tag{116}$$

A similar relation applies to any thermodynamic diagram. It will be recalled that an area on such a diagram is equal to the work performed for a cyclic process undergone by unit mass about that area. In this case we are not dealing with unit mass, but with the entire lower atmosphere, so that the area, A, is not simply equal nor proportional to the energy input Q_a.

After sunset, solar radiation need no longer be considered and the net loss of energy at the surface due to infrared radiation must be balanced by fluxes down from the air and up from the ground, the downward heat flux including that of latent heat associated with condensation at the surface, visible as dew if the surface itself cannot diffuse water into the ground at a sufficiently rapid rate. Since the lapse rate in the lowest layers in the early morning hours depends on many complex factors, it is not fruitful to attempt to forecast the appropriate area on a thermodynamic diagram comparable to A, above, nor to obtain the minimum temperature by such an approach. A direct solution of the heat conduction equations is possible with appropriate simplifications, giving a forecast of the minimum temperature at the ground (the most critical parameter for frost-damage assessment). If this temperature is expected to fall below the dew point of the air, fog formation (radiation fog) is very likely. The effect of the resultant condensation on air temperature and on liquid water content of the fog can be estimated by the procedures of (Chapter VII, Section 5), or by their analytical analogues (i.e., finding the isobaric wet-bulb temperature of supersaturated air rather than the adiabatic wet-bulb temperature). From (Chapter VII, Equation (29)),

$$T_w - T = \frac{l_v}{c_{pd}}(r - r_w), \qquad (117)$$

where T represents the expected minimum temperature (neglecting the possibility of condensation), T_w the actual minimum temperature (with fog containing a liquid water mixing ratio of $(r-r_w)$, r the saturation mixing ratio at the dew point temperature (we assume $T_d > T$) and r_w the saturation mixing ratio at the final temperature T_w. We may therefore state that

$$r - r_w = \left(\frac{\partial r_w}{\partial T}\right)_p (T_d - T_w). \qquad (118)$$

Introducing $r_w \simeq \varepsilon l_w/p$ and the Clausius-Clapeyron equation, Equation (118) becomes

$$r - r_w = \frac{\varepsilon r l_v}{R_d T_d^2}(T_d - T_w). \qquad (119)$$

From Equations (117) and (119),

$$(T_w - T_d) + (T_d - T) = \frac{\varepsilon l_v^2 r}{R_d c_{pd} T_d^2}(T_d - T_w).$$

Thus,

$$(T_d - T_w)\left(1 + \frac{\varepsilon r l_v^2}{c_{pd} R_d T_d^2}\right) = (T_d - T). \qquad (120)$$

From Equation (120), $(T_d - T_w)$ can be found, and then $(r - r_w)$, from Equation (119). An estimate of the horizontal visibility may be made from the following table:

$(r-r_w)$ in g kg^{-1}:	0.015	0.025	0.065	0.09	0.15	0.25	0.35	0.65	1.8
Visibility in m:	900	600	300	240	180	120	90	60	30

Once a radiation fog has formed, it will tend to thicken as long as radiational cooling of the fog plus ground can continue, since a dense fog acts as a black body as far as infrared radiation is concerned. Thus the major further cooling takes place at the fog top, and the maximum fog density may even be at some distance above ground level. After sunrise, a dense fog may continue to cool since the solar radiation is largely reflected and little absorbed by the fog. The ground itself will commence to warm slowly (as a result of the radiation transmitted through the fog), and the increased turbulence will initially mix the fog near the ground. At times this process may actually produce a further decrease in surface visibility shortly after sunrise. Eventually, however, the rising surface temperature and the net gain of radiational energy by the fog itself will lead to fog dissipation. For a thick fog, the fog will tend to dissipate first at the surface, leaving a low layer of stratus cloud above. This cloud becomes progressively thinner and less dense as the lower layers continue to warm, and disappears completely when the level of cloud top (earlier, the level of fog top) has the same potential temperature as the entire layer below.

Overcast cloud layers above the layer of diurnal temperature changes will normally exhibit a net cooling as a result of radiative processes, although there may be a slight gain of energy during the middle of the day. Since the cloud top acts as a black body to infrared radiation (but definitely not to solar radiation), the atmosphere immediately above the cloud behaves, for most if not all of the 24 h, like the air above the ground at night, except that the analogue to fog formation is of course a cloud-thickening process.

Finally, let us consider the consequences of radiative cooling of broken or scattered clouds, particularly when relatively thin. For a given cloud height, the radiative cooling rate is inversely proportional to cloud depth, to a first approximation. For a thin cloud, this cooling rate may far exceed that of the air between clouds, so that the cloud soon becomes denser than its environment and thus subsides, pseudo-adiabatically, until it reaches equilibrium with the environment at a lower level. This process normally continues until the liquid water content of the cloud is entirely depleted and the cloud has dissipated; this process is responsible for most cases of nocturnal cloud dissipation, e.g., for scattered or broken stratocumulus, etc., in the early evening. The net effect of the cooling plus subsidence is that the cloud descends along the environment curve, but a number of different cases may arise, depending on the ambient lapse rates and the cloud liquid water content.

Various possible cases will be illustrated by schematic tephigram curves. CS

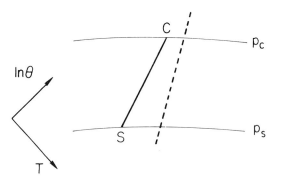

Fig. IX-21. Radiative cooling of scattered or broken clouds. Case I.

denotes the environment sounding between p_c, the initial cloud level, and p_s, the surface pressure; this segment will be considered to have a mean lapse rate of $\bar{\gamma}$. The saturation mixing ratio line denoted by short dashes refers to a value equal to the total water content of the cloud air, r_t (condensed phase plus vapor, which is a constant of the system); the corresponding lapse rate will be referred to as γ_r (its numerical value is of no concern here, since we are interested in graphical representation and analysis, but could be easily derived by differentiating $r_w = \varepsilon e_w/p = $ const. and introducing the Clausius-Clapeyron equation, $d\phi = v\, dp$ and the definition $\gamma_r = -dT/d\phi$; there results $\gamma_r = T/\varepsilon l_v$).

Case I, $\bar{\gamma} < \gamma_r$ (Figure IX-21). In this case the cloud does not dissipate but thickens as it descends $-r_w$ decreasing with r_t constant. At the surface there would be produced relatively dense fog patches. The effect of mixing with the environment during descent (probably a relative minimum with such a stable lapse rate) would tend to offset the cloud density increase.

Case II, $\bar{\gamma} = \gamma_r$ (Figure IX-22). In this case also, the cloud does not dissipate, but reaches the ground as fog patches, with no change in content of condensed phase (if mixing can be neglected).

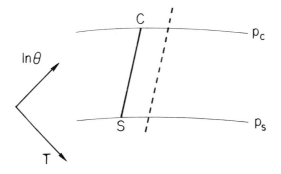

Fig. IX-22. Radiative cooling of scattered or broken clouds. Case II.

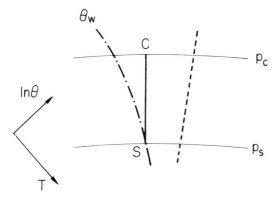

Fig. IX-23. Radiative cooling of scattered or broken clouds. Case IIIa.

Case IIIa, $\gamma_w > \bar{\gamma} > \gamma_r$ and $r_w(T_s) < r_t$ (Figure IX-23). In this case (for which a representative pseudo-adiabat is labeled as θ_w), the cloud does not dissipate completely as it sinks. Instead, the cloud density decreases and the cloud reaches the ground as thin fog patches. The increased turbulent mixing in this case, compared to cases I and II, could well lead to cloud dissipation by mixing with the environment air (especially if the latter is relatively dry).

Case IIIb, $\gamma_w > \bar{\gamma} > \gamma_r$ and $r_w(T_s) > r_t$ (Figure IX-24). In this case the cloud descends only to P, where $r_w = r_t$. The cloud has dissipated at this point and the descending volume is in equilibrium with the environment.

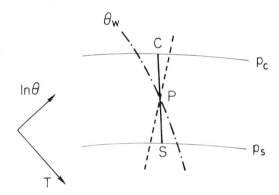

Fig. IX-24. Radiative cooling of scattered or broken clouds. Case IIIb.

If $\gamma_w > \bar{\gamma}$, the atmospheric stratification is stable for ascent or descent of saturated air, with or without a condensed phase. If $\gamma_d > \bar{\gamma} > \gamma_w$, the atmospheric stratification is stable for ascent or descent of unsaturated air, is unstable for ascent of saturated air, with or without condensed phase, and is unstable for descent of saturated air containing liquid water (or ice). Clouds produced by convection (cumulus) or by vertical mixing (stratocumulus) will have nearly dry-adiabatic lapse rates below the cloud, at least during the active formation stage, and in addition it is possible to have $\bar{\gamma} > \gamma_w$ below clouds produced by orographic ascent or by the general ascending motion associated with the stratified clouds in cyclonic weather systems. In general, the liquid water content at the base of all such clouds is small, so that descending motions triggered by instability soon lead to dissolution of such cloud elements (within the drier air below cloud base), but this does not contribute to dissipation of the cloud as a whole if active formation processes are simultaneously in operation. Radiative cooling of broken or scattered clouds under these conditions (which, we note, do not apply to overcast layers) will accelerate the dissipation of the entire cloud system, and this process is undoubtedly a potent one in the evening hours when the effects of convection and vertical mixing are greatly reduced, i.e., for cumulus and stratocumulus.

Case IVa, $\gamma_d > \bar{\gamma}\gamma_w$ and $\theta(P) > \theta(S)$ (Figure IX-25). With slight cooling, the cloud becomes unstable for descent and subsides rapidly to P, where it dissipates. The ex-cloud air at P is denser than the environment, and sinks dry-adiabatically to R, where it is in equilibrium with the environment.

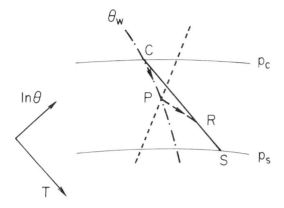

Fig. IX-25. Radiative cooling of scattered or broken clouds. Case IVa.

Case IVb, $\gamma_d > \bar{\gamma} > \gamma_w$ and $\theta(P) < \theta(S)$ (Figure IX-26). As before, the cloud cools slightly and descends rapidly along a pseudo-adiabat to P, where it dissipates completely. The ex-cloud air at P now sinks dry-adiabatically right down to the ground (at R), producing local cool down-drafts.

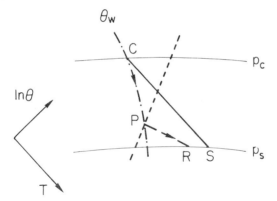

Fig. IX-26. Radiative cooling of scattered or broken clouds. Case IVb.

We may now compute the total amount of heat (Q_r) that must be lost by radiative processes in order that the cloud may be dissolved or reach the ground, whichever occurs first. During the first stage of the stepwise process, isobaric cooling accompanied by latent heat release, the first principle of thermodynamics (cf. Chapter IV, Equation (104)) gives, ignoring the heat content of the vapor and of the condensed phase,

$$\delta Q_r = c_{p_d} dT_1 + l_v dr_w = c_{p_d} dT_1 + l_v \left(\frac{\partial r_w}{\partial T}\right)_p dT_1. \tag{121}$$

During the second stage of the stepwise process, the reversible saturated adiabatic descent (the process analyzed in Chapter VII, Section 8), we may approximate the corresponding relation by neglecting the variation of l_v/T, giving

$$c_{p_d} dT_2 - R_d T\, d\ln p_2 + l_v\, dr_w = 0. \tag{122}$$

Expansion of dr_w in Equation (122) gives

$$c_{p_d} dT_2 + l_v \left(\frac{\partial r_w}{\partial T}\right)_p dT_2 - R_d T\, d\ln p_2 + l_v \left(\frac{\partial r_w}{\partial p}\right)_T dp_2 = 0. \tag{123}$$

Let us now add Equations (121) and (123), and introduce $\delta T = dT_1 + dT_2$ as the temperature change of the environment air for an infinitesimal pressure change, δp. Since $\delta p = dp_2$, it follows that

$$\delta Q_r = c_{p_d} \delta T + l_v \left\{ \left(\frac{\partial r_w}{\partial T}\right)_p \delta T + \left(\frac{\partial r_w}{\partial p}\right)_T \delta p \right\} - R_d T\, \delta \ln p. \tag{124}$$

Introducing the hydrostatic Equation (Chapter VIII, Equation (21)), and denoting by δr_w the change of saturation mixing ratio along the environment curve (for an increment $\delta\varphi$, of geopotential), we have

$$\delta Q_r = c_{p_d} \delta T + l_v\, \delta r_w + \delta \varphi. \tag{125}$$

This equation can now be integrated over the layer from C to S (or C to P, in case IIIb), treating l_v as a constant and computing the change in geopotential by the method of the mean adiabat (Chapter VIII, Section 11), i.e., from

$$\delta\varphi = -c_{p_d}\, \delta T_\theta. \tag{126}$$

The resulting value of Q_r is the heat which must be dissipated, per unit mass of the cloud. This should be converted to the heat loss per unit area of the entire cloud, to be consistent with the convention for radiative fluxes.

9.14. Maximum Rate of Precipitation

Let us consider now a unit mass of saturated air rising in the atmosphere. During the ascent, vapor will condense into water (or ice); by imagining that all the water precipitates as rain (or snow), we may compute an upper limit for the possible rate of precipitation. We shall proceed now to make this computation.

We start from Chapter VII, Equation (84) for the pseudoadiabats. In its last term, l_v can be treated as a constant without much error. We then develop

$$d\left(\frac{r_w}{T}\right) = \frac{dr_w}{T} - \frac{r_w}{T^2} dT,$$

multiply the whole equation by T and make the substitution $R_d T\, d\ln p = -d\phi$. We then obtain:

$$\left[c_{p_d} + \left(c_w - \frac{l_v}{T}\right)r_w\right] dT + d\phi + l_v\, dr_w = 0.$$

The second term within the square bracket usually amounts to some units per cent of c_{pd} (it reaches 10% at $r_w \cong 0.02$) and will be neglected in this approximate treatment. The equation can then be rearranged into:

$$-\frac{dr_w}{d\phi} = \frac{1 - (\gamma_w/\gamma_d)}{l_v} \qquad (127)$$

where Equation (12) has been taken into account, as well as the fact that $-dT/d\phi = \gamma_w$ in our case.

Equation (127) gives the condensed water per unit geopotential of ascent. We shall now assume that we have a saturated layer $\delta\phi$ thick, rising at a velocity U. The mass of air contained in this layer per unit area is $\varrho \, \delta z = \delta\phi/gv$.

We consider that the derivative with minus sign

$$-\frac{dr_w}{dt} = -\frac{dr_w}{d\phi}\frac{d\phi}{dt} \qquad (128)$$

expresses the mass of condensed water per unit mass of air and unit time. We have also:

$$\frac{d\phi}{dt} = g\frac{dz}{dt} = gU. \qquad (129)$$

Introducing this expression and Equation (127) in (128), and multiplying by the mass of air, we shall have the mass of water condensed (and eventually precipitated) per unit time and unit area:

$$P = \frac{1 - (\gamma_w/\gamma_d)}{l_v v} U \, \delta\phi \qquad (130)$$

P is here given, if MKS units are used, in kg m^{-2} s^{-1}. The mass of water precipitated per unit area may be expressed by the depth that it fills, in mm. Taking into account that the density is 10^3 kg m^{-3}, 1 kg m^{-2} of precipitated water is equivalent to 1 mm. If we further refer the precipitation to 1 h rather than to 1 s, we must multiply the rate of precipitation by 3600 in order to have it expressed in mm h^{-1}, as is common practice in meteorology:

$$P = 3600 \frac{1 - (\gamma_w/\gamma_d)}{l_v v} U \, \delta\phi \quad (\text{mm h}^{-1}) \qquad (131)$$

where all quantities are in MKS units.

The coefficient of $U \delta\phi$ in Equation (131) depends only on T and p, and so will P for a given value of U and $\delta\phi$. Therefore they will be defined for each point of a diagram. We can then choose, for instance, $U = 1$ m s^{-1} and $\delta\phi = 100$ gpm $= 981$ J kg^{-1}, and draw isopleths of constant P. If we are considering an ascending saturated layer at (T, p), we can read P on the diagram, and this will give us the maximum precipitation rate for each m s^{-1} of vertical velocity and each 100 gpm of thickness.

A similar computation may be performed for snow precipitation.

Figure IX-27 shows the shape of these isopleths on a tephigram. They are labeled with the value of P and 'S' or 'R' for 'snow' and 'rain', respectively.

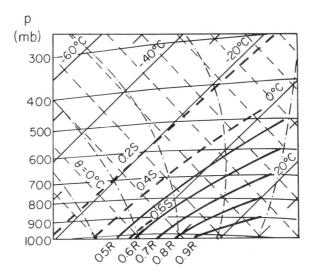

Fig. IX-27. Isopleths of maximum precipitation rate, on a tephigram. R: rain; S: snow.

Let us consider now a technique for the evaluation of the maximum rate of precipitation in terms of the distribution in the vertical of the vertical motion in isobaric coordinates, dp/dt. This turns out to be particularly straightforward on a tephigram, but could be adapted readily to any other thermodynamic diagram.

We may state, completely analogous to Equation (130), above,

$$P = \int \frac{dr_w}{dt} \frac{dp}{g}, \tag{132}$$

where the integration is taken over the entire cloud layer (or layers) to give the precipitation rate in mass of water per unit time per unit area.

Introducing Chapter IV, Equation (114) into Equation (122), we have

$$-\frac{dr_w}{dt} = \frac{T}{l_v} \frac{ds_d}{dt} \tag{133}$$

in terms of the dry-air entropy, s_d – a basic coordinate of the tephigram. The time derivative is that associated with the pseudo-adiabatic process, which implies that

$$\frac{ds_d}{dt} = \left(\frac{\partial s_d}{\partial p}\right)_{\theta_w} \frac{dp}{dt}. \tag{134}$$

Substituting Equations (133) and (134) into (132), we obtain the required integral, which may also be expressed as a summation over finite layers, e.g.,

$$P = -\int \frac{T}{gl_v} \frac{dp}{dt} \left(\frac{\partial s_d}{\partial p}\right)_{\theta_w} dp = \sum \left(\frac{T}{gl_v}\right)\left(\frac{dp}{dt}\right)(\Delta s_d)_{\theta_w}, \qquad (135)$$

where mean values are employed of T/l_v and of dp/dt for each layer, and where $(\Delta s_d)_{\theta_w}$ represents the dry-air entropy decrease along a mean pseudo-adiabat downward through the layer in question (we note this can be read off directly, on a tephigram, and is by definition a negative quantity).

In practice, isobaric vertical motions can be deduced from numerical weather prediction calculations, in either a diagnostic or prognostic mode (in the latter case, a prognostic sounding would also have to be employed, plus an indirect estimation of the depth of saturated layers). In theory, at least, we could employ observed rates of precipitation in order to verify the accuracy of computation of vertical motions. This is not a satisfactory procedure, however, since precipitation amounts are seldom representative, and a few rain gauges do not constitute an adequate sample of a large area for quantitative purposes. If instability phenomena are present, and their existence is not always obvious from synoptic weather data, the above equation will seriously underpredict precipitation, and in this latter instance horizontal variability will be large and random, increasing the difficulty of verification. Moreover, it will seldom be obvious from vertical soundings what was the precise vertical extent of the cloud layers (even if verification is being carried out with diagnostic vertical motions). Finally, the assumptions made (constant cloud properties, no evaporation below the cloud) could seldom be justified in individual cases. Thus, the equation for P is useful chiefly for semi-quantitative prediction for a large area.

9.15. Internal and Potential Energy in the Atmosphere

The Earth as a whole is in a state of radiative equilibrium, in which the solar radiation absorbed is compensated by emission to space, essentially as infrared radiation corresponding to surface and atmospheric temperatures.

Between absorption and emission, however, there is a complicated pattern of energy transformations. The heating by the Sun causes an increase in the specific internal energy of the atmospheric gas. As the gas expands due to the increase in temperature, it lifts the mass center of any vertical column that may be considered; this implies an increase in potential energy. There is also a partial transformation of the radiant energy received into vaporization enthalpy. The internal and potential energies may be partially transformed into kinetic energy of motion of large air masses, into turbulence energy, into mechanical work over the Earth's surface and finally into heat. The last conversion implies again an increase in internal and potential energy, thus closing a cycle. Other transformations between these different forms of energy are also possible. Differential latitudinal heating and horizontal transport are implied in this complex picture.

We have the following basic types of energy in the atmosphere: (1) potential (gravitational), (2) internal (thermal), and (3) kinetic. If we consider the approximate formulas Chapter IV, Equations (111) and (105):

$$u = c_v T + l_v q + \text{const.}$$

$$h = c_p T + l_v q + \text{const.}$$

we can see that the internal energy and the enthalpy will each include a term proportional, at every point, to the temperature, and another due to the vaporization heat of its water vapor content. It is customary to refer to the former as 'sensible heat', and to the latter as 'latent heat'. The kinetic energy can appear in large scale motions, in vertical convection or in a whole spectrum of eddies.

Here we shall only consider the expressions of potential and internal energy, their inter-relation and their possible transformation into kinetic energy of vertical motions. Moreover, we shall only consider in the internal energy and in the enthalpy the term proportional to the temperature, as we shall not study the more complex case when there are water phase transitions. We shall call U, P and K the internal, potential and kinetic energy, respectively, and H, as usual, the enthalpy.

Let us consider a vertical atmospheric column of unit cross section, in hydrostatic equilibrium, extending from the surface ($z=0$) to the height h. Each infinitesimal layer dz has a mass $\varrho\,dz$ and the internal energy $c_v T \varrho\,dz$*. We shall have for the column (introducing the hydrostatic equilibrium equation Chapter VIII, Equation (19)):

$$U = c_v \int_0^h T\varrho\,dz = \frac{c_v}{gR} \int_0^{\phi_h} p\,d\phi = \frac{c_v}{g} \int_{p_h}^{p_0} T\,dp \qquad (136)$$

where the small variations of c_v and R with humidity and of g with altitude are neglected.

The same integration can be performed for the enthalpy. It is easily seen that $H = \eta U$.

For the potential energy:

$$P = \int_0^h \phi\varrho\,dz = g\int_0^h \varrho z\,dz = \int_{p_h}^{p_0} z\,dp. \qquad (137)$$

Integrating by parts and introducing the gas law:

$$P = -p_h h + R\int_0^h \varrho T\,dz = -p_h h + \frac{R}{c_v}U = -p_h h + (\eta - 1)U. \qquad (138)$$

---

* We are taking for convenience the arbitrary additive constant as zero, which amounts to considering as reference state one obtained by extrapolation of the formula $u = c_v T$ to $T = 0$ K.

If the column extends to the top of the atmosphere, $p_h = 0$ and Equation (138) becomes

$$P = (\eta - 1) U \tag{139}$$

a proportionality which must always be obeyed, provided there is hydrostatic equilibrium. As $\eta = 7/5$, we have the relation

$$P : U = 2 : 5. \tag{140}$$

If the column receives heat from external sources, it must become distributed between the potential and the internal energy in the same proportion; i.e., $2/7 = 29\%$ must go to increase P and $5/7 = 71\%$ to increase U.

If we add both energies for such a column, we obtain

$$P + U = \eta U = H \tag{141}$$

where the last equation results by considering that ηU is given by expressions similar to Equation (136), but with c_p instead of c_v, in the coefficient. Thus, for a column in hydrostatic equilibrium extending in height to negligible pressures, the total enthalpy is equal to the sum of the internal and potential energies, and to η times the internal energy. Obviously, any energy received from external sources produces an equivalent increase in the enthalpy.

Now let us consider an infinite column divided, for the purpose of the discussion, into a lower part extending from ground to $z = h$ corresponding to a constant given P_h and with energies P, U, and an upper part, from $z = h$ upwards and with energies P', U' (see Figure IX-28). Let P_t, U_t be the values for the total column. The lower part must obey Equation (138). For the upper part, performing the same integration, but this time between the limits $z = h$ and $z = \infty$ (viz. from $p = p_h$ to $p = 0$), we arrive at

$$P' = p_h h + (\eta - 1) U'. \tag{142}$$

The sum of Equations (138) and (142) gives

$$P + P' = (\eta - 1)(U + U') \tag{143}$$

i.e.,

$$P_t = (\eta - 1) U_t \tag{144}$$

which is again the Equation (139), as applied to the present case. We notice now that for a process which does not alter the atmospheric structure above P_h (such as absorption of heat by the lower part), U' remains constant, while P' changes because of the change in h:

$$\Delta P' = p_h \Delta h \tag{145}$$

where p_h, which represents the weight of the whole column of unit cross section above the initial level h, must remain constant. That is, the change in P' is equal to the change in potential energy that would be experienced by a solid weight p_h (per unit cross section) resting on top of the lower part of the column, as this top is raised by Δh.

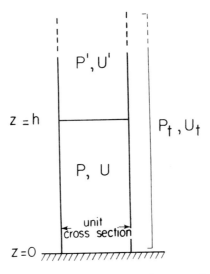

Fig. IX-28. Internal and potential energy of an atmospheric column.

From Equations (138), (144), and (145),

$$\Delta P_t = \Delta(P + p_h h) = (\eta - 1)\Delta U . \tag{146}$$

Summarizing, any energy input will go entirely to an increase $\Delta H_t = \Delta(P_t + U)$, from which the fraction 5/7 goes to ΔU and 2/7 to ΔP_t; from this latter portion, $p_h \Delta h$ gives the increase $\Delta P'$ and the rest goes to ΔP.

If we now consider an adiabatic process in which there are no motions with a component normal to the limiting surfaces of the system, and we neglect the work of external friction forces, we must have

$$\Delta(K + P_t + U) = 0 \tag{147}$$

$$\Delta K = -\Delta(P_t + U) = -\eta \Delta U = -\Delta H. \tag{148}$$

That is, the kinetic energy that can be produced is given by the decrease in the sum $H_t = (P_t + U)$ when passing from the initial to the final stratification. This is the maximum value attainable in principle, as no real process will be strictly adiabatic; there will be, in a larger or lesser degree, simultaneous reconversion of kinetic into internal and potential energy through frictional dissipation.

9.16. Internal and Potential Energy of a Layer with Constant Lapse Rate

Let us consider an atmospheric column in hydrostatic equilibrium with a thermal gradient $\gamma = $ const., extending from $z = 0$ to $z = h$. Using the equations, Chapter VIII, Equation (37) and Chapter VIII, Equation (40) to integrate the second expression (136) of

U, we obtain:

$$U = \frac{c_v}{gR} p_0 \int_0^{\phi_h} \left(1 - \frac{\gamma}{T_0}\phi\right)^{1/R\gamma} d\phi$$

$$= -\frac{c_v p_0 T_0}{gR\gamma} \left.\frac{\left(1 - \dfrac{\gamma}{T_0}\phi\right)^{(1/R\gamma)+1}}{\dfrac{1}{R\gamma}+1}\right|_0^{\phi_h} \quad (149)$$

$$= \frac{c_v}{g(1+R\gamma)}(p_0 T_0 - p_h T_h),$$

where real temperatures and gradients are used, instead of the virtual ones.

For the particular case of an adiabatic layer, $\gamma = 1/c_p$ and $R\gamma = \varkappa$. We then have

$$U = \frac{c_v}{g(1+\varkappa)}(p_0 T_0 - p_h T_h)$$

$$= \frac{c_p}{g(2\eta - 1)}(p_0 T_0 - p_h T_h). \quad (150)$$

And, writing the temperature in terms of the pressure and the potential temperature

$$T_0 = \theta\left(\frac{p_0}{p_{00}}\right)^{\varkappa}; \quad T_h = \theta\left(\frac{p_h}{p_{00}}\right)^{\varkappa}$$

we obtain

$$U = \frac{c_v \theta}{g(1+\varkappa)p_{00}^{\varkappa}}(p_0^{\varkappa+1} - p_h^{\varkappa+1}), \quad (151)$$

where $p_{00} = 1000$ mb. Therefore, for an adiabatic layer and given p_0 and p_h, the internal energy is proportional to its potential temperature, θ.

9.17. Margules' Calculations on Overturning of Air Masses

Margules used the previous formulas to calculate the change in potential and internal energy and to estimate the possible production of kinetic energy, in the following idealized case.

We shall assume two air layers of unit cross section, superimposed, both adiabatic, with the potential temperature of the upper layer θ_2 smaller than that of the lower layer θ_1, so that the system is unstable. p_0, p_m and p_h are the pressures at the bottom of the column, at the interface between the two layers and at the top, respectively (see Figure IX-29).

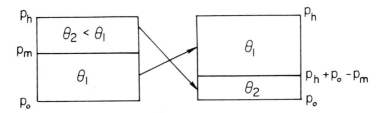

Fig. IX-29. Overturning of two adiabatic layers, according to Margules.

We assume now that an overturning takes place, by which the two layers exchange positions, but without any mixing taking place between the air masses. It is easy to see that the pressures at the bottom and at the top will be the same, but at the interface between the two layers it will now be $(p_h + p_0 - p_m)$, because this surface is holding the weight of the atmosphere above the whole column, given by p_h, and that of the layer θ_1, given (from the initial state) by $p_0 - p_m$.

We now apply Equation (151) to calculate the initial (U_i) and final (U_f) internal energies of the system

$$U_i = \frac{c_v}{g(1+\varkappa)p_{00}^\varkappa} [\theta_1(p_0^{\varkappa+1} - p_m^{\varkappa+1}) + \theta_2(p_m^{\varkappa+1} - p_h^{\varkappa+1})] \quad (152)$$

$$U_f = \frac{c_v}{g(1+\varkappa)p_{00}^\varkappa} \{\theta_1[(p_h + p_0 - p_m)^{\varkappa+1} - p_h^{\varkappa+1}] + \\ + \theta_2[p_0^{\varkappa+1} - (p_h + p_0 - p_m)^{\varkappa+1}]\}. \quad (153)$$

The variation of U in the overturning will therefore be:

$$\Delta U = \frac{c_v}{g(1+\varkappa)p_{00}^\varkappa} (\theta_1 - \theta_2) \times \\ \times [(p_h + p_0 - p_m)^{\varkappa+1} + p_m^{\varkappa+1} - p_0^{\varkappa+1} - p_h^{\varkappa+1}], \quad (154)$$

which can also be written:

$$\Delta U = -\frac{c_v}{g(1+\varkappa)} \left(\frac{p_0}{p_{00}}\right)^\varkappa (\theta_1 - \theta_2) \left[p_0 + p_h\left(\frac{p_h}{p_0}\right)^\varkappa - p_m\left(\frac{p_m}{p_0}\right)^\varkappa - \right. \\ \left. - (p_0 + p_h - p_m)\left(\frac{p_0 + p_h - p_m}{p_0}\right)^\varkappa\right]. \quad (155)$$

In the case of layers of moderate thickness, so that p_0 and p_h do not differ greatly, the following approximations can be made in Equation (155):

$$\left(\frac{p}{p_0}\right)^\varkappa = \left(1 + \frac{p-p_0}{p_0}\right)^\varkappa = 1 + \varkappa\frac{p-p_0}{p_0} + \ldots \cong 1 + \varkappa\frac{p-p_0}{p_0}.$$

Introducing this simplification, Equation (155) becomes:

$$\Delta U \cong -\frac{2a}{g}\left(\frac{p_{00}}{p_0}\right)^{1-\varkappa}(\theta_1 - \theta_2)\frac{(p_0 - p_m)(p_m - p_h)}{p_{00}} \quad (156)$$

with

$$a = \frac{c_v \varkappa}{1 + \varkappa} = \frac{R}{\eta(1 + \varkappa)} = \frac{R}{2\eta - 1}.$$

Equation (156) shows that the change in internal energy will be proportional to the difference of potential temperatures and to the product of the pressure thicknesses of the two layers. The same proportionality will hold, according to Equations (146) and (148), for ΔP_t and for ΔK, with the appropriate change of coefficients. In particular:

$$\Delta K \cong \frac{2a\eta}{g}\left(\frac{p_{00}}{p_0}\right)^{1-\varkappa}(\theta_1 - \theta_2)\frac{(p_0 - p_m)(p_m - p_h)}{p_{00}}. \quad (157)$$

This is the maximum value of kinetic energy that can be produced, in principle, during the overturning; if the total system is isolated, this kinetic energy will finally dissipate again into internal and potential energy (with changes in θ_1, θ_2 or both). We can use the value obtained in Equation (157) to compute the velocity W related to it by

$$\Delta K = \tfrac{1}{2}MW^2 \quad (158)$$

where $M = (p_0 - p_h)/g$ is the total mass (per unit cross section). As this is such an idealized model, we are only interested in the resulting order of magnitude. For layers 100 mb thick, with potential temperatures differing by 10 °C, we obtain $W = 21$ m s^{-1}; with 200 mb, $W = 30$ m s^{-1}. This is the order of magnitude of the strongest updraughts in storms, but we must remember that in real storms there is condensation, the updraughts are localized over certain areas, and the whole process is much more complex and implies a great deal of turbulent mixing.

Computation shows that the values of ΔK are only a small fraction of U. For instance, with $\theta_1 = 300$ K, $\theta_2 = 290$ K and 100 mb thick layers starting at 1000 mb, ΔK is about 920 times (and ΔU about 1290 times) smaller than $U = 4.2 \times 10^8$ J m^{-2}.

Margules also calculated the variation of potential and internal energy for the case of adjacent layers, as shown in Figure IX-30, and obtained similar results.

Fig. IX-30. Overturning of adjacent adiabatic layers.

9.18. Transformations of a Layer with Constant Lapse Rate

Let us consider a layer extending from $z=0$ to $z=h$, with a constant lapse rate which in general will be different from γ_d. Its internal energy will be given by Equation (149).

We shall only consider qualitatively the processes that may occur in this layer.

(a) If vertical mixing takes place, the resulting potential temperature will be uniform and given by the weighted average θ through all the layer (cf. Chapter VII, Section 12). The initial lapse rate is γ and the final one is the adiabatic lapse rate γ_d. It may be shown that according to whether $\gamma \gtreqless \gamma_d$, the variation of internal energy will be $\Delta U \lesseqgtr 0$. Thus, if the layer had a superadiabatic lapse rate, the mixing may occur spontaneously, with a decrease in potential and internal energy, and the layer will acquire an adiabatic lapse rate. In this case, there will be a surplus of energy which, assuming that the process occurs without exchange with the environment, will be first transformed into kinetic energy and then dissipated into heat by turbulence. The final potential temperature of the layer will be such as to maintain the initial value of U and P. If $\gamma = \gamma_d$, the layer is in neutral equilibrium with respect to vertical exchanges.

If $\gamma < \gamma_d$, the layer is stable; in order to produce the vertical mixing, which occurs with an increase in potential and internal energy, the layer must absorb energy from external sources.

(b) Littwin considered another ideal process in a layer with uniform superadiabatic lapse rate γ_i: the total orderly overturning of the layer, in such a way that the stratification is inverted, with the highest layers passing to the lowest positions and viceversa. The final gradient γ_f is subadiabatic. During the overturning, each infinitesimal layer follows a dry adiabat to its new location. It is assumed that no mixing occurs. The process occurs with decrease in potential and internal energy. Figure IX-31 illustrates this case; it may be noticed that a discontinuity has been assumed at p_h, to indicate that the upper part (stable), does not participate in the process.

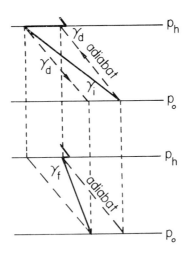

Fig. IX-31. Overturning of a layer with a superadiabatic constant lapse rate.

As Margules did, Littwin computed the kinetic energy corresponding to the decrease in $(P_t + U)$, and from it, the mean velocities attainable during the overturning, with results of the same order as of the Margules calculation, i.e. similar to our results based on Equation (158).

Obviously Littwin's process, although more elaborate than Margules', still is an unrealistic idealized model. Particularly, the condition of absence of mixing will not hold in the atmosphere. Real processes will be intermediate between (a) and (b), and in particular for local convection they may be closer to case (a); actually we have seen that convection caused by ground heating during strong insolation leads to thorough mixing and an adiabatic lapse rate in the lowest layers.

9.19. The Available Potential Energy

As we have already remarked, the pioneering calculations of Margules were aimed at understanding the source of the energy of storms and based on crude models, rather inadequate when considered as a local representation of strong convective processes. But the same concepts have been carried over to large-scale motions, mainly by Lorenz, where they find fruitful application and can be related to the problems of the general circulation and of the energy conversions in the atmosphere. We shall not develop his theory, which would be beyond the scope of this book, but shall mention briefly the basic ideas and indicate how the concepts of the previous sections can be generalized.

Again we shall restrict consideration to processes in which condensation plays no role (or its role is minor and can be ignored) and shall assume that conditions of vertical hydrostatic equilibrium are essentially valid everywhere. Equations (136) to (141) are therefore applicable over each particular location. If we now integrate the expressions of U and P for columns extending to the top of the atmosphere (i.e., to $p = 0$) over an extended surface or over the whole surface of the earth, we shall have the total values of U and P for that extended region or for the whole atmosphere, respectively. It is customary, in the context of the present problem to call the sum $(P + U)$ the *total potential energy*. It must be stressed, however, that as Equation (141) is valid at every location, it is also valid for the integrated value over the whole surface under consideration; therefore the total potential energy is identically the *total enthalpy* $H = P + U$.

We shall define *adiabatic motion* (or more generally *adiabatic process*) as a motion (or process) in which entropy, and therefore potential temperature, is conserved for every parcel of air.

Atmospheric motion is not in general adiabatic; friction over the ground surface or in the air and mixing are nonadiabatic processes. However, friction is the only nonadiabatic process which directly alters the kinetic energy, destroying it and generating internal energy. The remaining nonadiabatic processes alter only the internal energy directly. So the only source of kinetic energy K in the atmosphere is $H = P + U$.

The advantage of considering adiabatic flows resides in that, in spite of the previous remarks, they represent a good approximation to large-scale motions in the atmosphere.

In an adiabatic process, an air parcel moves on an isentropic surface, i.e., a constant θ

surface. If the parcel is to be accelerated, there must be a pressure gradient on the surface; without it, the kinetic energy of the parcel will remain constant. Therefore, in a situation where constant p surfaces and constant θ surfaces coincide, there can be no conversion of H into K, no matter how large H is. The question that becomes important then is what is the portion of H that can be converted into K. This is indeed a very difficult question.

An easier question, however, can be asked. For a given state of the atmosphere, what is the minimum value of the total enthalpy (or total potential energy) that can be attained by an adiabatic redistribution of mass in the atmosphere? We call this minimum value H_{min}, and we define now the *available potential energy* A as the following difference:

$$A = H - H_{min}. \tag{159}$$

It will be noticed that we are considering now for the whole atmosphere (or a large portion of it) a process similar to the local adiabatic overturnings of Margules and Littwin from an initial, unstable stratification to a final stable one, where U and H have minimum values.

The advantage of defining A in this manner lies in that it separates the part relevant to the production of air motions from the much larger bulk of total enthalpy, mostly unavailable for conversion into kinetic energy. The convenience of doing so can be appreciated in the following two examples: (a) in an atmosphere with horizontal, hydrostatically stable stratification, however large H may be, $A = 0$; (b) on the other hand, removing heat differentially from this atmosphere will decrease H, but will create $A > 0$, causing instability (i.e., will decrease H_{min} more than it does H).

It is clear that the available potential energy has the following properties:
(1) $(A + K)$ is conserved under adiabatic flow.
(2) A is completely determined by the distribution of mass.
(3) $A = 0$ if the stratification is such that constant p and constant θ coincide everywhere.
(4) $A > 0$ if constant p and constant θ surfaces do not coincide.

According to property (1), A can be considered as the only source of kinetic energy:

$$\Delta(A + K) = 0. \tag{160}$$

Equation (160) is the generalized equivalent of our previous Equation (147). It is important to recognize that, for a given situation, not all of A may be converted into K in the real atmosphere, because the redistribution of mass required to achieve H_{min} may not satisfy the equation of motion. The available potential energy A merely gives an upper bound for the enthalpy to kinetic energy conversion. In fact, it can be estimated that typically only about 1/10 of the total enthalpy of the atmosphere is transformed into kinetic energy. As the available potential energy is found to be of the order of 1/200 of the total enthalpy, only about 1/2000 of the latter transforms into K.

Equation (147) indicates that A is the only sink for K in adiabatic flow, but, as mentioned before, in real flows friction will dissipate part of the kinetic energy, producing an increase in H_{min} but not in A.

Now let us derive an explicit expression for A.

Since θ increases monotonically with height (according to the assumed stable vertical hydrostatic equilibrium), we shall use θ instead of p or z as the vertical coordinate. $p(x, y, \theta)$ may be considered as the weight of air with potential temperature exceeding θ, at x, y. This is true even if $\theta < \theta_0$, where θ_0 is the surface potential temperature (i.e., extending formally the range of the variable θ_0 to underground locations), provided that we define $p(x, y, \theta) = p_0(x, y)$, where p_0 is the surface pressure, for $\theta < \theta_0$.

The average of p over an isentropic surface of area S is

$$\bar{p}(\theta) = \frac{1}{S} \int_S p(x, y, \theta) \, dS \tag{161}$$

$\bar{p}$ is conserved under adiabatic redistribution of mass, because $S\bar{p}$ gives the total weight of air with potential temperature exceeding θ, which is conserved.

The total enthalpy for the area S is

$$H = \frac{c_p}{g} \int_S \int_0^{p_0} T \, dp \, dS = \frac{c_p}{g p_{00}^\varkappa} \int_S \int_0^{p_0} \theta p^\varkappa \, dp \, dS$$

$$= \frac{c_p}{g(1 + \varkappa) p_{00}^\varkappa} \int_S \left[\int_0^{p_0} \theta \, dp^{1+\varkappa} \right] dS \tag{162}$$

where $p_{00} = 1000$ mb and c_p and g are considered as constants. Let us discuss the integral between the bracket; solving by parts:

$$\int_0^{p_0} \theta \, dp^{1+\varkappa} = \theta p^{1+\varkappa} \Big|_0^{p_0} - \int_{\theta_0}^{\infty} p^{1+\varkappa} \, d\theta$$

$$= \theta_0 p_0^{1+\varkappa} - \int_{\theta_0}^{\infty} p^{1+\varkappa} \, d\theta \tag{163}$$

where the lower limit value of the first integral at the right-hand side has been set equal to 0 because p decreases with height more rapidly (essentially as an exponential variation) than θ increases. Now, according to the condition mentioned above for the extension of the range of θ, p remains constant and equal to p_0 for $\theta < \theta_0$, so that

$$\theta_0 p_0^{1+\varkappa} = \int_0^{\theta} p^{1+\varkappa} \, d\theta \tag{164}$$

and (163) becomes

$$\int_0^{p_0} \theta \, dp^{1+\varkappa} = \int_0^{\infty} p^{1+\varkappa} \, d\theta. \tag{165}$$

Introducing (165) into (162):

$$H = \frac{c_p}{g(1+\varkappa)p_{00}^{\varkappa}} \int_S \int_0^{\infty} p^{1+\varkappa} \, d\theta \, dS. \tag{166}$$

Now $H_{\min}$ can be achieved by rearranging mass in such a way that p is constant on the isentropic surfaces, and this constant p should be equal to the earlier defined $\bar{p}$, because $\bar{p}$ is conserved under adiabatic processes. Therefore

$$H_{\min} = \frac{c_p}{g(1+\varkappa)p_{00}^{\varkappa}} \int_S \int_0^{\infty} \bar{p}^{1+\varkappa} \, d\theta \, dS \tag{167}$$

and

$$A = \frac{c_p}{g(1+\varkappa)p_{00}^{\varkappa}} \int_S \int_0^{\infty} (p^{1+\varkappa} - \bar{p}^{1+\varkappa}) \, d\theta \, dS. \tag{168}$$

We can write

$$p = \bar{p} + p' \tag{169}$$

where usually $p' \ll p$; this allows us to make the approximation

$$p^{1+\varkappa} = \bar{p}^{1+\varkappa}\left(1 + \frac{p'}{\bar{p}}\right)^{1+\varkappa} \cong \bar{p}^{1+\varkappa} \times$$

$$\times \left[1 + (1+\varkappa)\frac{p'}{\bar{p}} + \frac{(1+\varkappa)\varkappa}{2}\left(\frac{p'}{\bar{p}}\right)^2\right] \tag{170}$$

which, introduced into (168), gives

$$A = \frac{c_p}{g(1+\varkappa)p_{00}^{\varkappa}} \left[(1+\varkappa) \int_S \int_0^{\infty} \bar{p}^{1+\varkappa} \frac{p'}{\bar{p}} \, d\theta \, dS + \right.$$

$$\left. + \frac{(1+\varkappa)\varkappa}{2} \int_S \int_0^{\infty} \bar{p}^{1+\varkappa}\left(\frac{p'}{\bar{p}}\right)^2 d\theta \, dS\right]. \tag{171}$$

The first integral within the bracket vanishes, as can be readily seen by replacing p' from (169), integrating first over the surface ($\bar{p}$ being a constant for this integration) and considering (161). Equation (171) reduces therefore to

$$A = \frac{c_p \varkappa}{2gp_{00}^{\varkappa}} \int_S \int_0^\infty \bar{p}^{1+\varkappa} \left(\frac{p'}{\bar{p}}\right)^2 d\theta \, dS. \tag{172}$$

If we want to express A in terms of temperature, we write

$$T = \bar{T} + T' \tag{173}$$

where $\bar{T}$, the average temperature over an isentropic surface, is defined in a similar way to $\bar{p}$ (Equation (161)) and $T' \ll T$. On each isentropic surface, we have the relations:

$$T = \theta \left(\frac{p}{p_{00}}\right)^{\varkappa} = \theta \left(\frac{\bar{p}}{p_{00}}\right)^{\varkappa} \left(1 + \frac{p'}{\bar{p}}\right)^{\varkappa} \cong \bar{T}\left(1 + \varkappa \frac{p'}{p}\right). \tag{174}$$

Therefore

$$\frac{p'}{\bar{p}} = \frac{1}{\varkappa} \frac{T'}{\bar{T}} \tag{175}$$

which, introduced into (172), gives finally

$$A = \frac{c_p}{2g\varkappa p_{00}^{\varkappa}} \int_S \int_0^\infty \bar{p}^{1+\varkappa} \left(\frac{T'}{\bar{T}}\right)^2 d\theta \, dS. \tag{176}$$

Let us consider an example based on a simplified model. We assume a rectangular surface of length L and width W. We take the variable x along the length and y along the width. All parameters will be assumed independent of y.

Figure IX-32 shows a plot of constant p surfaces in the x, θ plane. With respect to the

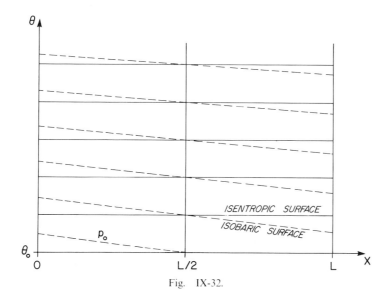

Fig. IX-32.

horizontal isentropic surfaces, the isobaric surfaces are assumed to be inclined with a constant slope, according to the linear variation

$$p(x, \theta) = p\left(\frac{L}{2}, \theta\right)\left[1 + s\left(1 - \frac{2x}{L}\right)\right] \qquad (177)$$

where s is a constant determining the slope. We further assume that the lapse rate is constant throughout the atmosphere at $x = L/2$, with a value γ. The air is assumed dry for convenience (or humidity corrections are neglected). The solution of the example is left to Problem 18. Here we shall only give the results, assuming the following parameters. The surface pressure at the midpoint is $p_0(L/2, \theta_0) = 1000$ mb, the surface potential temperature is $\theta_0 = 300$, K, $\gamma = 6.5$ K gpkm^{-1} and $s = 0.15$ (which means that the pressures on the isentropic surface θ_0 vary up to ± 150 mb). The calculation shows that the available potential energy and the total enthalpy, referred to unit area, are:

$$A = 3.56 \times 10^6 \text{ J m}^{-2}, \qquad H = 2.58 \times 10^9 \text{ J m}^{-2}$$

and

$$A/H = 1/725.$$

PROBLEMS

1. Prove that the work of expansion of an air parcel ascending adiabatically and quasi-statically may be expressed by the following formula:

$$\delta a = -\frac{gT}{\eta T_e} dz$$

where T_e is the temperature of the environment, $\eta = c_p/c_v$, and a is referred to unit mass.

2. From the following sounding

p (mb)	T (°C)	U_w %	p (mb)	T (°C)	U_w %
920	24.0	68	568	− 1.2	42
900	22.5	70	545	− 2.0	49
850	20.0	83	500	− 8.0	77
800	15.8	83	400	− 19.5	71
765	13.0	92	300	− 33.0	
735	12.8	55	250	− 41.5	
700	10.0	54	200	− 54.0	
645	5.8	55	150	− 65.5	
600	3.0	32			

(a) Compute the positive and negative instability areas for a surface parcel, on a tephigram.

(b) Compute the same for a parcel at 600 mb.

(c) Determine the thickness (base and top pressures) of the layer with latent instability.

(d) Determine the lifting condensation level (LCL) and the free convection level (FCL) for a surface parcel.

(e) Compute the instability index, defined as the temperature of a parcel from 850 mb when taken to the 500 mb level minus the temperature of the atmosphere at the same level.

(f) Assuming that after the time of the sounding and due to insolation, the lower layer absorbs heat from the ground and cumuli start developing, what will be the level of the cloud base?

3. Given the following temperature sounding

p (mb)	T (°C)
950	22.5
900	18.0
850	15.0
800	16.0
750	12.0
700	7.0
650	4.0
600	− 1.5
500	− 10.0
400	− 20.0

and knowing that the dew-point at the surface (950 mb) is 15.7 °C, plot the sounding on a tephigram and

(a) Determine, using the tephigram, the mixing ratio r, the relative humidity U_w, the potential temperature θ, the wet-bulb temperature T_{aw}, the potential wet-bulb temperature θ_{aw} and the potential equivalent temperature θ_{ae} for a surface parcel. Mark the relevant points on the diagram. Compute the isobaric wet-bulb temperature T_{iw} and equivalent temperature T_{ie}.

(b) Determine the lifting condensation level and the free convection level for a surface parcel. What can you say of the conditional instability of surface parcels?

4. Plot the following points on an aerological diagram:

p (mb)	T (°C)	r (g kg^{-1})
1000	17.0	10.0
900	–	9.0
850	8.0	6.0
800	–	3.8
700	− 4.5	–
500	− 20.0	–
300	− 35.0	–

Join them by straight lines for each variable, and assume that the two representations obtained correspond to an atmospheric sounding. Find the potential temperature θ, the wet-bulb potential temperature θ_w, the lifting condensation level (LCL), the saturation temperature T_s and the free convection level (FCL) for a ground (1000 mb) parcel, and mark on the diagram the negative and positive areas of instability for vertical parcel displacement. Find the convective condensation level (CCL). Indicate the layer with latent instability.

5. The vertical distribution of temperature and humidity (mixing ratio r) of the atmosphere over a certain location is given initially by the following data:

p (mb)	$T(°C)$	r (g kg^{-1})
1000	20.0	11.5
850	12.0	9.0
700	2.0	5.0
600	− 5.5	2.5
500	− 14.5	1.5

Radiative heating of the ground results in the development of convection, attainment of the convective condensation level (CCL) and cumuli formation. Using a tephigram, find:

(a) The lifting condensation level (LCL) for a ground parcel before the heating, and the CCL.

(b) The dew point, the pseudo-wet bulb temperature and the potential pseudo-wet bulb temperature for ground parcels before and after the heating.

(c) The approximate vertical velocity acquired by an air parcel in a cumulus at 600 mb, as predicted by the parcel theory. Assume that the initial velocity at the CCL is negligible and use the area equivalence given for the diagram.

(d) Indicate the layer with latent instability at the initial time (before the heating).

6. Given the following data:

p(mb): 1000 900 800 700 600 500 400
$T(°C)$: 25   18   10   2 −4.5 −12 −20

and knowing that the relative humidity at the ground (1000 mb) is 60% and that the mixing ratio has an average value of 4 g kg^{-1} between 600 and 800 mb;

(a) determine, on a tephigram, the values of the following parameters for ground parcels: adiabatic potential wet bulb temperature θ_{aw}, dew point temperature T_d, lifting condensation level (LCL) and free convection level (FCL).

(b) What type of conditional instability do these parcels have?

(c) On a separate tephigram, compute the thickness of the layer between 800 and 600 mb, in a single step, both by the method of mean temperature and of the mean adiabat. (Use the tephigram as if it were a skew emagram.) Express the results in gpm and in m^2 s^{-2}.

7. The potential pseudo-wet bulb temperature decreases with height through an atmospheric layer ($\delta\theta_{aw}/\delta z < 0$). What comments can you make on its vertical stability, if the layer is saturated? And if it is not saturated?
8. An unsaturated air mass is ascending through the isobar p. Due to radiant heat exchange, the virtual potential temperature increases by 2.8 K gpkm^{-1}.
   Determine the state of stability of the atmosphere at that level if its (geometric) virtual temperature lapse rate is 7.3 K gpkm, and
   (a) if $p = 800$ mb;
   (b) if $p = 500$ mb.
9. A vertically ascending air particle is receiving some heat. The temperature of the particle is increasing by 0.0134 K m^{-1}. Determine
   (a) the polytropic exponent n and
   (b) the heat flux.
10. A parcel of unsaturated air is receiving heat from the surroundings through heat conduction a rate of $\delta q/dz$ during its ascent. The virtual lapse rate of the atmosphere is 5 K gpkm^{-1}. Determine the stability of the atmosphere for this situation, when
    (a) $\delta q/dz = 2$ cal kg^{-1} m^{-1}
    (b) $\delta q/dz = 1$ cal kg^{-1} m^{-1}.
11. An air particle, warmer than the surroundings by 10 K, is moving upwards. Doing so it receives some heat due to absorption of long wave radiation. How far will the particle move, if the temperature lapse rate in the atmosphere is 6 K gpkm^{-1} and the particle receives heat according to the polytropic law $pv^n = $ const, with $n = 1.53$?
12. An air mass is ascending. The initial pressure is 1000 mb and the initial temperature is 290 K. Due to radiant heat exchange its potential temperature is increasing at the rate of 3.45% per kilometer. How much heat per unit mass does the air receive in the first 100 m of ascent?
13. A saturated layer 300 m thick is ascending at 2 m s^{-1} at the level of 850 mb. Its mean temperature is 20 °C. What is the maximum rate of precipitation that can be expected from it?
14. Consider a column of atmospheric air extending upwards to negligible pressures with a constant lapse rate $\gamma = 5$ K gpkm^{-1}. The pressure and temperature at the ground are: $p_0 = 1000$ mb, $T_0 = 300$ K. Compute its total internal energy, potential energy and enthalpy, per unit cross section. Neglect variation in the acceleration of gravity ($g = 9.8$ m s^{-2}); the air is unsaturated everywhere and can be considered as dry air.
15. Consider an adiabatic layer of dry air, extending from 500 to 400 mb. The temperature at the base is 0 °C.
    (a) What is the thickness of that layer, in gpm?
    (b) Compute the differences in specific internal energy u, enthalpy h, and entropy s, between parcels at the base and at the top.
    (c) What is the total internal energy of a column of that layer with unit cross section?

16. The vertical structure of a column of dry air is given by the following points:

   1000 mb (ground) $-15.0\,°C$
   900 mb $-10.0\,°C$
   800 mb $-13.0\,°C$.

   After a certain time, this column has changed to:

   1000 mb (ground) $+5.0\,°C$
   900 mb $-2.0\,°C$
   800 mb $-7.0\,°C$.

   The atmosphere above 800 mb is assumed to preserve its temperature stratification during the process.
   (a) Compute with a diagram the initial and final thicknesses, in gpm, by any graphical method.
   (b) How much have the isobars of $\leqslant 800$ mb been lifted?
   (c) Compute the changes in internal energy and in potential energy (per unit cross section), and the total amount of energy that must have been absorbed by the column, in order to undergo this change.

17. In Section 17 some numerical results were given for the overturning of two 100 mb-thick layers, starting at 1000 mb, with potential temperatures 300 K and 290 K, according to Margules. Check the values obtained for U, ΔU, ΔK and W.

18. At the end of Section 19 a numerical example was given for a simple situation. Perform, for that example, the calculation of the total potential energy of enthalpy H and available potential energy A.

APPENDIX I

TABLE OF PHYSICAL CONSTANTS

Molecular weights.

$M_d = 28.964$ (g mol^{-1}); $M_v = 18.015$ (g mol^{-1})

Gas constants

$R^* = 8.314$ J mol^{-1}K^{-1}; $\quad \varepsilon = R_d/R_v = 0.622$
$R_d = 287.05$ J kg^{-1}K^{-1}; $\quad R_v = 461.51$ J kg^{-1}K^{-1}

Specific heat capacities

Dry air $\quad c_{p_d} = 1005$ J kg^{-1}K^{-1} = 0.240 cal g^{-1}K^{-1}
(see also $\quad c_{v_d} = \;\;718$ J kg^{-1}K^{-1} = 0.171 cal g^{-1}K^{-1}
Table II-1) $\quad \varkappa_d = 0.286; \quad \eta_d = 1.40$

Water vapor $\quad c_{p_v} = 1850$ J kg^{-1}K^{-1} = 0.443 cal g^{-1}K^{-1}
$\quad c_{v_v} = 1390$ J kg^{-1}K^{-1} = 0.332 cal g^{-1}K^{-1}

Water, at 0°C $\quad c_w = 4218$ J kg^{-1}K^{-1} = 1.008 cal g^{-1}K^{-1}
(see other values
in table below)

Ice, at 0°C $\quad c_i = 2106$ J kg^{-1}K^{-1} = 0.503 cal g^{-1}K^{-1}

Critical and triple point constants of water: see Chapter IV, Section 6.

Latent heats of phase transformations of water, at 0°C (see other values in table below)

$l_f = 0.334 \times 10^6$ J kg^{-1} = $\;\;$79.7 cal g^{-1}
$l_v = 2.501 \times 10^6$ J kg^{-1} = 597.3 cal g^{-1}
$l_s = 2.834 \times 10^6$ J kg^{-1} = 677.0 cal g^{-1}

Thermodynamic properties of condensed water (see also Chapter IV, Table IV-5)

t °C	c_i J kg^{-1} K^{-1}	e_i mb	l_s 10^6 J kg^{-1}	l_f 10^6 J kg^{-1}	l_v 10^6 J kg^{-1}	e_w mb	c_w J kg^{-1} K^{-1}
−100	1382	1.402×10^{-5}	2.824				
−90	1449	9.665×10^{-5}	2.828				
−80	1520	5.468×10^{-4}	2.832				
−70	1591	2.614×10^{-3}	2.834				
−60	1662	1.080×10^{-2}	2.837				
−50	1738	3.933×10^{-2}	2.8383	0.2035	2.6348	0.06354	5400
−40	1813	0.1283	2.8387	0.2357	2.6030	0.1891	4770
−30	1884	0.3797	2.8387	0.2638	2.5749	0.5087	4520
−20	1959	1.032	2.8383	0.2889	2.5494	1.254	4350
−10	2031	2.597	2.8366	0.3119	2.5247	2.862	4270
0	2106	6.106	2.8345	0.3337	2.5008	6.107	4218
5					2.4891	8.718	4202
10					2.4774	12.27	4192
15					2.4656	17.04	4186
20					2.4535	23.37	4182
25					2.4418	31.67	4180
30					2.4300	42.43	4179
35					2.4183	56.23	4178
40					2.4062	73.77	4178
45					2.3945	95.85	4179
50					2.3823	123.39	4181

APPENDIX I

Saturation vapor pressures over pure liquid water (e_w) and over pure ice (e_i) as functions of temperature – Detailed table.

$T(°C)$	e_w (mb)	e_i (mb)	T	e_w	e_i	T	e_w	T	e_w
−50	0.0635	0.0393	−24	0.8826	0.6983	1	6.565	26	33.606
−49	0.0712	0.0445	−23	0.9647	0.7708	2	7.054	27	35.646
−48	0.0797	0.0502	−22	1.0536	0.8501	3	7.574	28	37.793
−47	0.0892	0.0567	−21	1.1498	0.9366	4	8.128	29	40.052
−46	0.0996	0.0639	−20	1.2538	1.032	5	8.718	30	42.427
−45	0.1111	0.0720	−19	1.3661	1.135	6	9.345	31	44.924
−44	0.1230	0.0810	−18	1.4874	1.248	7	10.012	32	47.548
−43	0.1379	0.0910	−17	1.6183	1.371	8	10.720	33	50.303
−42	0.1533	0.1021	−16	1.7594	1.505	9	11.473	34	53.197
−41	0.1704	0.1145	−15	1.9114	1.651	10	12.271	35	56.233
−40	0.1891	0.1283	−14	2.0751	1.810	11	13.118	36	59.418
−39	0.2097	0.1436	−13	2.2512	1.983	12	14.016	37	62.759
−38	0.2322	0.1606	−12	2.4405	2.171	13	14.967	38	66.260
−37	0.2570	0.1794	−11	2.6438	2.375	14	15.975	39	69.930
−36	0.2841	0.2002	−10	2.8622	2.597	15	17.042	40	73.773
−35	0.3138	0.2232	−9	3.0965	2.837	16	18.171	41	77.798
−34	0.3463	0.2487	−8	3.3478	3.097	17	19.365	42	82.011
−33	0.3817	0.2768	−7	3.6171	3.379	18	20.628	43	86.419
−32	0.4204	0.3078	−6	3.9055	3.684	19	21.962	44	91.029
−31	0.4627	0.3420	−5	4.2142	4.014	20	23.371	45	95.850
−30	0.5087	0.3797	−4	4.5444	4.371	21	24.858	46	100.89
−29	0.5588	0.4212	−3	4.8974	4.756	22	26.428	47	106.15
−28	0.6133	0.4668	−2	5.2745	5.173	23	28.083	48	111.65
−27	0.6726	0.5169	−1	5.6772	5.622	24	29.829	49	117.40
−26	0.7369	0.5719	0	6.1070	6.106	25	31.668	50	123.39
−25	0.8068	0.6322	–	–	–	–	–	–	–

Saturation mixing ratio with respect to water.

Saturation (with respect to water) adiabatic lapse rate.

t °C	r_w (g kg^{-1})				γ_w (K gpkm^{-1})	
	1000 mb	800 mb	600 mb	400 mb	1000 mb	500 mb
−40	0.118	0.148	0.197	0.295	9.78	9.54
−30	0.318	0.397	0.529	0.794	9.41	8.90
−20	0.785	0.980	1.31	1.96	8.79	7.85
−10	1.79	2.24	2.99	4.49	7.80	6.50
0	3.84	4.80	6.41	9.66	6.59	5.17
10	7.76	9.72	13.02	19.73	5.39	4.11
20	14.95	18.79	25.29	38.69	4.40	3.35
30	27.69	34.98	47.50	74.02	3.66	2.87
40	49.81	63.49	87.56	141.2	3.15	2.55

Surface tension of water against air

$t = -10\,°C, \quad \sigma = 0.07729\ \text{Nm}^{-1}$
$t = 0\,°C, \quad \sigma = 0.07570\ \text{Nm}^{-1}$
$t = 20\,°C, \quad \sigma = 0.07275\ \text{Nm}^{-1}$

Lapse rates $\gamma_d = 9.76\ \text{K gpkm}^{-1}$; γ_w: see table above.

Gravity

Normal acceleration of gravity $g_0 = 9.80665\ \text{m s}^{-2}$
(other values = see Chapter VIII, Section 1)
Gravitational constant $G = 6.67 \times 10^{-11}\ \text{Nm}^2\ \text{kg}^{-2}$
Mass of earth $M_e = 5.98 \times 10^{24}\ \text{kg}$
Angular velocity of earth $\omega = 7.292 \times 10^{-5}\ \text{s}^{-1}$

Standard atmosphere: see Chapter VIII, Table VIII-1

Conversion factors

1 atm = 1.01325×10^5 Pa
1 mb = 10^2 Pa = 10^3 dyn cm^{-2}
1 IT cal = 4.1868 J = 4.1868×10^7 erg
1 gpm = 9.80665 J kg^{-1} = 0.980665 dyn-m

BIBLIOGRAPHY

Very few books have been devoted specifically to the Thermodynamics of the Atmosphere, and some are old and unavailable. The subject is, however, treated, although with little development in general, in many textbooks devoted to General and Dynamic Meteorology, or in comprehensive compendia or handbooks. We have collected a bibliography on atmospheric thermodynamics and related subjects, classified according to the type of publication, hoping that this may give some orientation for further reading. Alphabetical order within each section has been followed. No claim is made for exhaustiveness.

1. *Monographs on Thermodynamics of the Atmosphere*

Dufour, L. and van Mieghem, J.: 1975, *Thermodynamique de l'Atmosphère*, Institut Royal Météorologique de Belgique, 278 pp. This book appears to be an extension of the earlier version by van Mieghem and Dufour (see below).
Raethjen, P.: 1942, *Statik und Thermodynamik der Atmosphäre*, Berlin.
Stüve, G.: 1937, Thermodynamik der Atmosphäre, In *Handbuch der Geophysik*, Vol 9, Berlin, Gebr. Borntraeger, pp. 173–314.
van Mieghem, J. and Dufour, L.: 1949, *Thermodynamique de l'atmosphère*, Office International de Librairie, Bruxelles, 247 pp.
Wegener, A.: 1911, Thermodynamik der Atmosphäre, J. A. Barth, Leipzig, 331 pp.

2. *Articles or Chapters in Encyclopedic Works*

Berry, F. A., Bollay, E., and Beers, N. R. (eds.): 1945, *Handbook of Meteorology*, McGraw-Hill, 1068 pp. Section V (pp. 313–409) by N. R. Beers is an excellent summary on Meteorological Thermodynamics and Atmospheric Statics. Physics of Atmospheric Phenomena and Radiation are among the other sections.
Godske, C. L., Bergeron, T., Bjerkness, J, and Bundgaard, R. C.: 1957, *Dynamic Meteorology and Weather Forecasting*. Publ. jointly by Amer. Met. Soc., Boston and Carnegie Institution, Washington, 800 pp. Part I of this work (119 pp.) is an excellent treatment of the Thermodynamics and Statics of the Atmosphere, which includes Thermodynamics (Ch. 1), Statics, with a detailed discussion of vertical stability (Ch. 2), Quasi-Static Diagnosis of Atmospheric Fields (Ch. 3) and Radiation Processes (Ch. 4). It contains an extensive bibliography at the end of each chapter.
Linke, F.: 1939, *Meteorologisches Taschenbuch*, Vol. IV, Akad. Verlagsgesellschaft, 286 pp.
Malone, T. F. (ed.): 1951, *Compendium of Meteorology*, Amer. Met. Soc., Boston, 1334 pp. The only thermodynamics included in this voluminous work is an article by *J. van Mieghem* on 'Application of the Thermodynamics of Open Systems to Meteorology' (pp. 531–538) and a brief summary by *F. Möller* on 'Thermodynamics of Clouds' (pp. 199–206).
Shaw, N.: 1930, *Manual of Meteorology*, Vol. III, Cambridge, Univ. Press, Ch. VI–Air as Worker (pp. 205–268).

3. *Books on General or Dynamic Meteorology or other Subjects, Containing Atmospheric Thermodynamics*

Only some textbooks of a rather long list are mentioned here.

Belinskii, V. A.: 1948, *Dynamic Meteorology*, OGIZ, Moscow, Leningrad; Engl. transl.: The Israel Program for Scientific Translations, 1961. 591 pp.
Brunt, D.: 1952, *Physical and Dynamical Meteorology*. Cambridge, 428 pp. Chapters 2, 3 and 4 (77 pp.).
Byers, H. R.: 1974, *General Meteorology*. 4th edn., McGraw-Hill, 461 pp. The first six chapters (140 pp.).
Eskinazi, S.: 1975, *Fluid Mechanics and Thermodynamics of our Environment*, Academic Press, 1975. 422 pp.

Haltiner, G. J. and Martin, F. L.: 1957, *Dynamical and Physical Meteorology*, McGraw-Hill, 470 pp.
Haurwitz, B.: 1941, *Dynamic Meteorology*, McGraw-Hill, 365 pp. First four chapters (84 pp.).
Hess, S. L.: *Introduction to Theoretical Meteorology*; 1959, Holt, Rinehart and Winston, 362 pp. The first seven chapters (113 pp.).
Holmboe, J. Forsythe, G. E., and Gustin, W.: 1945, *Dynamic Meteorology*, Wiley and Chapman and Hall, 378 pp. The first five chapters (144 pp.).
Petterssen, S.: 1956, *Weather Analysis and Forecasting*. 2nd. ed., McGraw-Hill In Vol. II (Weather and Weather Systems) the emphasis is on the thermodynamics of atmospheric processes, treated in close connection with meteorological aspects.
Wegener, A. and Wegener, K.: 1935, *Physik der Atmosphäre*, J. A. Barth Verlag, Leipzig, 482 pp. Chapters 3 and 4.

4. General Thermodynamics

There are, of course, many texts on this general field of physics. Here we suggest only a few among the many that may be appropriate for further consultation. In particular, the thermodynamics of open systems (cf. Ch. IV) will be found developed in textbooks on chemical thermodynamics, rather than in standard general textbooks.

Glasstone, S.: 1947, *Thermodynamics for Chemists*, Van Nostrand, 522 pp.
Guggenheim, E. A.: 1949, *Thermodynamics*, North-Holland Publ. Co., 394 pp.
Kirkwood, J. G. and Oppenheim, I.: 1961, *Chemical Thermodynamics*, McGraw-Hill, 261 pp.
Zemansky, M. W.: 1957, *Heat and Thermodynamics*, McGraw-Hill, 658 pp.

5. General Description of the Atmosphere

Chamberlain, J. W.: 1978, *Theory of Planetary Atmospheres*, Academic Press, 330 pp.
Dobson, G. M. B.: 1968, *Exploring the Atmosphere*, 2nd edn., Clarendon, Oxford, 209 pp.
Fleagle, R. G. and Businger, J. A.: 1963, *Introduction to Atmospheric Physics*, Academic Press, 346 pp.
Goody, R. H. and Walker, J. C. G.: 1972, *Atmospheres*, Prentice Hall, 150 pp.
Iribarne, J. V. and Cho, H.-R.: 1980, *Atmospheric Physics*, D. Reidel, Dordrecht, 212 pp.
Wallace, J. M. and Hobbs, P. V.: 1977, *Atmospheric Science*, Academic Press, 466 pp.

6. Special Topics

The titles of the publications are self-explanatory. The list includes a short selection of classical papers.

Defrise, P., Godson, W. L., and Pône, R.: 1949, *Les Diagrammes aérologiques*, World Meteorological Organization, Techn. Publ. No. 66. TP. 25. A summary on aerological diagrams by P. Defrise is included as an Appendix in Van Mieghem and Dufour, *Thermodynamique de l'atmosphère*, 1949.
Laikhtman, D. L., Gandin, L. S., Danovich, A. M., Melnikova, I. I., Roozin, M. I., Sopotsko, E. A., and Shlenyova, M. V.: 1970, *Problems in Dynamic Meteorology*, World Meteorological Organization, Techn. Note No. 261. TP. 146, 245 pp. A collection of problems, including problems on atmospheric thermodynamics (Chapters 2 and 3) and statics (Ch. 5).
Lorenz, E. N.: 1955, Available Potential Energy and the Maintenance of the General Circulation, *Tellus* **7**, 157-167.
Margules, M.: 1905, 'Über die Energie der Stürme', *Zentr. Anst. Meteor. Wien* **40**, 1-26.
Margules, M.: 1906, 'Zur Sturmtheorie', *Meteor. Z.* **23**, 480-497.
Mason, B. J.: 1971, *The Physics of Clouds*, 2nd edn., Clarendon, Oxford, 671 pp. For further reading in the field where Chapter V of the present book finds application.
Normand, C. W. B.: 1938a, 'On the Instability from Water Vapour', *Quart. J. Roy. Met. Soc.* **64**, 47-69.
Normand, C. W. B.: 1938b, 'Kinetic Energy Liberated in an Unstable Layer; *Quart. J. Roy. Met. Soc.* **64**, 71-74.
Refsdal, A.: 1933, 'Zur Thermodynamik der Atmosphäre', *Meteor. Z.* **50**, 212-218.
Rossby, C. G.: 1932, *Thermodynamics Applied to Air Mass Analysis*, Meteor. Papers, Massachussetts Inst. Tech., Vol. 1, No. 3, 41 pp.
Schnaidt, F.: 1943, 'Über die adiabatischen Zuständsänderungen feuchter Luft, die abgeleiteten Temperaturen und den Energievorrat Atmosphärischer Schichtungen', *Geol. Beitr. Geoph.* **60**, 16-133.

Thomson, W. (Lord Kelvin): 1871, 'On the Equilibrium of Vapour at a Curved Surface of Liquid', *Phil. Trans. Roy. Soc. London*, **42**, 448–452.

7. Meteorological Tables

Beers, N. R.: 1945, 'Numerical and Graphical Data', in F. A. Berry, E. Bollay and N. R. Beers (eds), *Handbook of Meteorology*, McGraw-Hill. Section I, 121 pp.

Manual of the ICAO Standard Atmosphere, 2nd edn., International Civil Aviation Organization, Montreal, 1964, Doc. 7488/2, 182 pp.

Smithsonian Meteorological Tables (prep. by R. J. List), 6th edn., Smithsonian Institution, Washington, 1966, 527 pp.

World Meteorological Organization – *International Meteorological Tables*. W.M.O. – No. 188.TP.94, 1966 (plus 1968 and 1973 supplements).

8. Thermodynamic Properties of Water Vapor and of Air

Gibson, M. R. and Bruges, E. A.: 1967, 'New Equations for the Thermodynamic Properties of Saturated Water in both the Liquid and Vapour Phases', *J. Mech. Eng. Sci.* **9**, 24–35.

Goff, J. A.: 1949, 'Final Report of the Working Subcommittee of the International Joint Committee on Psychrometric Data', *Amer. Soc. Mech. Eng., Trans.* **71**, 903–913.

Keenan, J. H. and Keyes, F. G.: 1936, *Thermodynamic Properties of Steam*, Includes Data for the Liquid and Solid Phases, Wiley, 89 pp.

Keenan, J. H. and Kaye, J.: 1945, *Thermodynamic Properties of Air*, Wiley, 73 pp. Includes Polytropic Functions.

Kiefer, P. J.: 1941, 'The Thermodynamic Properties of Water Vapour', *Monthly Weather Rev.* **69**, 329–331.

ANSWERS TO PROBLEMS

Chapter I

1. $t' = 12°$. **2.** (a) $1\,\mathrm{m\,s^{-2}} = 10^2\,\mathrm{cm\,s^{-2}}$; $1\,\mathrm{kg\,m^{-3}} = 10^{-3}\,\mathrm{g\,cm^{-3}}$; $1\,\mathrm{N} = 10^5\,\mathrm{dyn}$; $1\,\mathrm{Pa} = 10\,\mu\mathrm{bar}$; $1\,\mathrm{J} = 10^7\,\mathrm{erg}$; $1\,\mathrm{J\,kg^{-1}} = 10^4\,\mathrm{erg\,g^{-1}}$. **3.** 17 atm. **5.** $\bar{M} = 28.90\,\mathrm{g\,mol^{-1}}$ $\bar{R} = 287.68\,\mathrm{J\,kg^{-1}\,K^{-1}}$.

Chapter II

1. (a) $A = nR^*T \ln(p_f/p_i)$ (n = number of moles); (b) $A = p_f(V_i - V_f) = V_i(p_f - p_i)$; (c) $A = nC_pT[(p_f/p_i)^\varkappa - 1]$; (d) $A = nR^*T(p_f/p_i - 1) = V_i(p_f - p_i) = p_f(V_i - V_f)$. $Q = -A$ in all processes. **2.** (a) $v_i = 0.813\,\mathrm{m^3\,kg^{-1}}$. (b) $T_f = 256\,\mathrm{K}$; $v_f = 1.049\,\mathrm{m^3\,kg^{-1}}$. (c) $\Delta u = -2.0 \times 10^4\,\mathrm{J\,kg^{-1}}$; $\Delta h = -2.8 \times 10^4\,\mathrm{J\,kg^{-1}}$. (d) $-A = 2.4 \times 10^{13}\,\mathrm{J}$. (e) The same as in (c) and 0, respectively. (f) $v_i = 0.59\,\mathrm{m^3\,kg^{-1}}$; $T_f = 245\,\mathrm{K}$; $v_f = 0.73\,\mathrm{m^3\,kg^{-1}}$; $\Delta u = -1.2 \times 10^4\,\mathrm{J\,kg^{-1}}$; $\Delta h = -2.0 \times 10^4\,\mathrm{J\,kg^{-1}}$. **3.** (a) $T_{Br} = 685\,\mathrm{K}$. (b) $A_B = 230\,\mathrm{J}$. (c) $T_{Ar} = 4777\,\mathrm{K}$; (d) $Q_A = 653\,\mathrm{cal}$. **4.** (a) $n = 1.262$. (b) $T_f = 205\,\mathrm{K}$. (c) $\Delta U = -1421\,\mathrm{J}$. (d) $A = -2168\,\mathrm{J}$. (e) $Q = 178\,\mathrm{cal}$. **5.** (a) $\Delta u = 574\,\mathrm{cal\,g^{-1}}$. (b) $\Delta h = 606\,\mathrm{cal\,g^{-1}}$. (c) $c_{pv} = 0.44\,\mathrm{cal\,g^{-1}\,K^{-1}}$.

Chapter III

1. $\Delta S = 5.76\,\mathrm{J\,K^{-1}}$; $\Delta G = -1729\,\mathrm{J}$. **2.** $\int_i^f \delta Q/T = \Delta S = -nR^* \ln(V_i/V_f)$ for processes (a), (c), (d). (b) $\int_i^f \delta Q/T = -p_f(V_i - V_f)/T < \Delta S$. **3.** $\Delta U = \Delta H = -747\,\mathrm{cal}$; $\Delta S = -2.72\,\mathrm{cal\,K^{-1}}$; $\Delta G = -28\,\mathrm{cal}$. At the triple point. **4.** (a) $DS \leq 0$. (b) $DG \geq VDp$; $DF \geq -pDV$. (c) $DG \geq 0$. **5.** $\Delta h = 4.1 \times 10^{-4}\,\mathrm{cal\,g^{-1}}$; $\Delta u = 6.9 \times 10^{-6}\,\mathrm{cal\,g^{-1}}$. **6.** (a) $\Delta h = -1.23 \times 10^{-2}\,\mathrm{cal\,g^{-1}}$; $\Delta s = -7.3 \times 10^{-7}\,\mathrm{cal\,g^{-1}\,K^{-1}}$. (b) $\Delta h = -10.1\,\mathrm{cal\,g^{-1}}$; $\Delta s = -3.8 \times 10^{-2}\,\mathrm{cal\,g^{-1}\,K^{-1}}$. **7.** $+11\%$. **8.** It decreases. **9.** $p = 300.2\,\mathrm{mb}$; $\theta = 331.5\,\mathrm{K}$. **10.** $dG < -S\,dT + V\,dp = 0$. **11.** (iii), which corresponds to reversible freezing at $0\,°\mathrm{C}$.

Chapter IV

1. $0.695\,\mathrm{atm}$. **2.** $592\,\mathrm{cal\,g^{-1}}$. **3.** $-7.5 \times 10^{-3}\,\mathrm{K\,atm^{-1}}$. **4.** (a) $\Delta S = 39.4\,\mathrm{cal\,mol^{-1}\,K^{-1}}$. (b) $-4.1\,\mathrm{cal\,mol^{-1}\,K^{-1}}$. **5.** $80\,\mathrm{cal\,g^{-1}}$. **6.** $e = 21.2\,\mathrm{mb}$; $r = 0.0135$; $q = 0.0133$; $c_p = 0.243\,\mathrm{cal\,K^{-1}\,g^{-1}}$; $T_v = 32.5\,°\mathrm{C}$; $\varkappa = 0.285$; $\theta = 30\,°\mathrm{C}$, $\theta_v = 32.5\,°\mathrm{C}$, both constant for adiabatic expansion. **7.** $T_v = 2.6\,°\mathrm{C}$; $R = 287.5\,\mathrm{J\,kg^{-1}\,K^{-1}}$; $c_p = 1007\,\mathrm{J\,kg^{-1}\,K^{-1}}$; $\varkappa = 0.286$. **8.** (a) $e = 8.6\,\mathrm{mb}$. (b) $r = 5.4\,\mathrm{g\,kg^{-1}}$. (c) $e = 6.9\,\mathrm{mb}$; r:

the same as before. **9.** 7.7 g kg^{-1}. **10.** 1.10. **11.** (1) $\Delta h = -11.3$ cal g^{-1}; $\Delta s =$ $= -0.040$ cal g^{-1} K^{-1}. (2) (a) $+0.7\%$ and $+0.7\%$, respectively; (b) -2.6% and -2.6%; (c) -1.4% and -1.4%. **12.** We assume thermal and mechanical equilibrium. Therefore T and p are uniform for all phases.

In order to determine the composition of each phase, we must know the molar ratios of all components minus one (the last one being determined by difference). This gives a total of $\varphi(c-1)$ variables of composition, where φ is the number of phases and c the number of components.

The total number of variables (T, p and composition variables) is then

$$2 + \varphi(c-1).$$

The values of these variables are restricted by the conditions of chemical equilibrium, i.e., that the chemical potential of each component must be equal in all phases:

$$\mu_{i1} = \mu_{i2} = \ldots = \mu_{i\varphi} \quad (i = 1, 2, \ldots c).$$

This gives $(\varphi-1)$ equations for each of the c values of i; therefore, a total of $c(\varphi-1)$ conditions.

The number of independent variables or *variance* v of the system will be the difference between the total number of variables and the restrictive conditions:

$$v = 2 + \varphi(c-1) - c(\varphi-1) = 2 + c - \varphi.$$

Chapter V

1. (a) 0.12 μm; (b) 1.20 μm; (c) 2.40 μm. **2.** (a) 1.9 μm; (b) 3.3 μm. **3.** (a) 31.7 mb; (b) 0.19%.

Chapter VI

1. After drawing the basic grid, notice that the intersections $\theta = T$ define $p = 1000$ mb. Apply the congruency property to draw the other isobars. Saturation mixing ratio lines are defined by $p = \varepsilon e_w(T)/r_w + e_w(T)$. **3.** $J = -c_p/p^* \neq$ const. **4.** $r = 11.3$ g kg^{-1}; $r_w = 16.2$ g kg^{-1}; $\theta = 28.5\,°C$; $T_d = 14.2\,°C$. **5.** $\Delta u = -7.17 \times 10^3$ J kg^{-1}; $\Delta h = -1.005 \times 10^4$ J kg^{-1}; $\Delta s = 10.2$ J kg^{-1} K^{-1}; $a = -1.00 \times 10^4$ J kg^{-1}; $q = 2.85 \times 10^3$ J kg^{-1}.

Chapter VII

1. $e_i = 15.8$ mb; $e_f = 11.9$ mb; $U_{w_i} = 68\%$; $U_{w_f} = 99\%$; $T_{d_f} = 9.8\,°C$. **2.** $T_f = 8.1\,°C$; $\Delta e_w = -1.5$ mb; $\Delta c = 1.2$ g m^{-3}. **3.** At every instant, $m_t = m_v$.

$$(m_d c_{pd} + m_v c_w)\,dT + d[l_v(T)m_v] = 0.$$

When integrating, the variation of m_v has to be taken into account:

$$\frac{dT}{l_v(T)} + \frac{dm_v}{m_d c_{p_d} + m_v c_{p_v}} = 0.$$

Introducing Kirchhoff's law: $dl_v(T) = (c_{p_v} - c_w) dT$, we have

$$\frac{1}{c_{p_v} - c_w} d \ln l_v(T) + \frac{1}{c_{p_v}} d \ln (m_d c_{p_d} + m_v c_{p_v}) = 0.$$

Integration gives the desired expression. **4.** $e' - e = -(c_p/\varepsilon l_v)p(T' - T)$; $T_{iw} = 3.3\,°C$; $T_{ie} = 15.4\,°C$. **5.** $T = 10.6\,°C$; $r = 8.1\,g\,kg^{-1}$; it is saturated and the liquid water content is $0.9\,g\,m^{-3}$. **6.** $1.3\,g\,m^{-3}$. **7.** Yes (cf. Equation (78)). **8.** Use formula (85) starting from a chosen point ($p = 1000$ mb and $T = 250, 270, 290$ for the three curves) to fix the constant. **9.** 286 mb. **10.** $-79\,°C$. **11.** The process occurs descending along the vapor pressure curve, say from an initial point A to a final point B (both on the curve). The second path consists of: (a) a line from A to an intermediate point C at the left of the curve; for moderate variations of T this line is virtually straight and its slope is as for the arrow starting from P in Figure VII-11; (b) a straight line going from C to B, with a slope as that of the straight line in Figure VII-6. **12.** (a) $R\Delta T (n - 1)$; (b) -1.44×10^4 J kg^{-1}; 7.2×10^3 J kg^{-1}. **13.** $T = [T_0/(1 - \varkappa)] [(p_1^{1-\varkappa} - p_2^{1-\varkappa})/(p_1 - p_2)]p^\varkappa$. **14.** $T_{aw} = 16.2\,°C$; $\theta_{aw} = 20.2\,°C$; $T_s = 13.0\,°C$; $p_s = 827$ mb. **15.** (a) $\theta_{ae} = \theta \exp (l_v r/c_p T_s) = T(p_0/p)^\varkappa \exp(l_v r/c_p T_s)$ where $p_0 = 1000$ mb, $r = r_w(T_s)$. Consider ascent from p, T first to saturation level p_s, T_s; then apply Equation (82) from $p_s T_s$ to $p'T'$ at a high enough level to consider $r'_w(T') \cong 0$; then apply equation to the descent from $p'T'$ to p_0, $T = \theta_{ae}$. **16.** $r_w = 5.0$ g kg^{-1}; $\theta = 10.8\,°C$; $\theta_v = 11.3\,°C$; $p_s = 810$ mb; $T_s = -6.0\,°C$; $T_{aw} = -0.8\,°C$; $\theta_{aw} = 4.4\,°C$; $T_d = -4.7\,°C$; $T_{iw} = -0.4\,°C$; $T_{ie} = 9.6\,°C$. **17.** It could be the air at 1000 mb; it could not be the air at 800 mb. **18.** $\theta_{aw}, \theta_{ae}; T_d, q, r$.

Chapter VIII

1. 18.8 K gpkm^{-1}. **2.** (a) $p_0 = 1004.8$ mb. (b) Differentiate $p_0 = p(1 + \phi \gamma_v/T_v)^{1/R_d\gamma_v}$ logarithmically, and make use of the approximations $\ln(1+x) \cong x - x^2/2$ and $(1+x)^{-1} \cong 1 - x$ for $x \ll 1$. The result is: $d \ln p_0 \cong -(\phi^2/2R_d T_v^2) d\gamma_v$, defining the relative error in p_0. (c) $dp_0 = 0.4$ mb. **3.** 1574 gpm. **4.** 842 gpm $= 826$ dyn-m $= 8260$ J kg^{-1}. **5.** Differentiate $\phi = R_d \bar{T} \ln (p_0/p_h)$ with respect to the horizontal distance x and obtain $\ln (p_0/p_h) = -(\bar{T}/p_0)(dp_0/dx)/(d\bar{T}/dx) = 0.1197$. Introduce into ϕ and find $z = \phi/g = 953$ m.

Chapter IX

1. By introducing the equations of hydrostatic equilibrium $dp = -(g/v_e) dz$ and of adiabatic expansion $pv^n = $ const. into $\delta a = -p\, dv$. **2.** (a) Negative: 0.020 cal g^{-1}; positive: 0.285 cal g^{-1}. (b) Both 0. (c) 920 to 750 mb. (d) LCL: 845 mb; FCL: 700 mb. (e) $I = 5.2\,°C$. (f) CCL: 780 mb. **3.** (a) $r = 12.0$ g kg^{-1}; $U_w = 67\%$; $\theta = 27.0\,°C$; $T_{aw} = 17.9\,°C$; $\theta_{aw} = 19.9\,°C$; $\theta_{ae} = 62\,°C$; $T_{iw} = 18.0\,°C$; $T_{ie} = 52.5\,°C$. (b) LCL: 860 mb;

FCL: 633 mb; pseudolatent type. **4.** $\theta = 17.0\,°C$; $\theta_w = 15.0\,°C$; LCL: 954 mb; $T_s = 13.1\,°C$; FCL: 870 mb; CCL: 910 mb; layer with latent instability: 1000 to 825 mb. **5.** (a) LCL: 945 mb; CCL: 835 mb. (b) Before: $T_d = 15.8\,°C$; $T_{aw} = \theta_{aw} = 17.2\,°C$. After: $T_d = 13.9\,°C$; $T_{aw} = \theta_{aw} = 18.2\,°C$. (c) $16.2\,\text{m s}^{-1}$. (d) 1000 to 800 mb. **6.** (a) $\theta_{aw} = 19.3\,°C$; $T_d = 16.7\,°C$; LCL = 890 mb; FCL = 830 mb. (b) real latent type. (c) 2324 gpm = $22791\,\text{m}^2\,\text{s}^{-2}$. **7.** If the layer is saturated, it is unstable with respect to parcel convection. If it is not saturated, it will be stable or unstable with respect to parcel convection, according to whether $\delta\theta/\delta z \gtrless 0$, and it is potentially unstable (regarding layer instability). **8.** (a) $\delta\theta_v/\delta\phi = 2.6\,\text{K gpkm}^{-1} < d\theta_v/d\phi = 2.8\,\text{K gpkm}^{-1}$: unstable. (b) stable. **9.** (a) $n = 1/(1 - R_d\beta/g) = 0.72$ ($\beta = -0.0134\,\text{K m}^{-1}$). (b) $\delta q/dz = g - c_p\beta = 5.6\,\text{cal kg}^{-1}$. **10.** Parcel lapse rate: $\gamma' = \gamma_d(1 - \delta q/dz)$. (a) $\gamma' = 1.4\,\text{K gpkm}^{-1} < 5\,\text{K gpkm}^{-1}$: unstable. (b) $\gamma' = 5.6\,\text{K gpkm}^{-1}$: stable. **11.** 1.72 km. **12.** $\delta q/dz = c_p(T/\theta) d\theta/dz$; 240 cal kg^{-1}. **13.** $5.2\,\text{mm h}^{-1}$. **14.** $U = 1.9 \times 10^9\,\text{J m}^{-2}$; $P = 0.77 \times 10^9\,\text{J m}^{-2}$; $H = 2.7 \times 10^9\,\text{J m}^{-2}$. **15.** (a) 1.73×10^3 gpm. (b) $\Delta h = 1.7 \times 10^4\,\text{J kg}^{-1}$; $\Delta u = 1.2 \times 10^4\,\text{J kg}^{-1}$; $\Delta s = 0$. (c) $U = 1.94 \times 10^8\,\text{J m}^{-2}$. **16.** (a) $\Delta\phi_i = 1706$ gpm; $\Delta\phi_f = 1773$ gpm. (b) 67 gpm. (c) $\Delta U = 1.56 \times 10^7\,\text{J m}^{-2}$; $\Delta P = 8.4 \times 10^5\,\text{J m}^{-2}$; $\Delta H = 2.18 \times 10^7\,\text{J m}^{-2}$.
18. From Equations (169) and (177),

$$\frac{p'}{p} = s\left(1 - \frac{2x}{L}\right)$$

From Chapter VIII, Equation (39) and expressing the temperatures in terms of potential temperature, we find for a constant lapse rate column

$$\frac{p}{p_0} = \left(\frac{\theta}{\theta_0}\right)^{-1/(\varkappa - R\gamma)}$$

In the present example, for $\theta > \theta_0$,

$$\bar{p}(\theta) = p_0\left(\frac{\theta}{\theta_0}\right)^{-1/(\varkappa - R\gamma)}$$

where $p_0 = p(L/2, \theta_0)$ and γ is the lapse rate at $x = L/2$. For $\theta < \theta_0$, $\bar{p}(\theta) = p_0$, as explained in the text. Applying Equation (172), we obtain

$$A = WL\frac{c_p\varkappa(1 + \varkappa)s^2}{6gp_{00}^\varkappa(1 + R\gamma)}p_0^{1+\varkappa}\theta_0$$

The total enthalpy H is calculated from Equation (149) with $p_h = 0$ and $H = \eta U$:

$$H = WL\frac{c_p\theta_0 p_0^{1+\varkappa}}{g(1 + R\gamma)p_{00}^\varkappa}$$

The values of A and H referred to unit area are obtained by dividing by WL. Introducing the values of p_0, θ_0, γ and s given in the text the numerical results mentioned are obtained.

INDEX OF SUBJECTS

Activation of hygroscopic nuclei 96
Adiabatic
   ascent 136, 138
   expansion 102, 136, 138
   isobaric processes 123
      mixing 127, 129
   motion 235
   process 235
   processes in ideal gases 28
   walls 3
Adiabats 29, 103
Aerological diagrams 98
Air
   atmospheric 12
   composition 12
   dry 13
   moist 69
Aircraft trails 130
Altimer 168
   setting 169
Altitude computation 172
Arctic Sea Smoke 130
Area
   computation and energy integrals 110
   equivalence 99, 107
Atmosphere
   constant lapse rate 163
   dry-adiabatic 165
   enthalpy 228
   homogeneous 164
   ICAO 167
   internal energy 227
   isothermal 166
   potential energy 227
   standard 166
Atmospheric statics 156
Available potential energy 235

Bubbles 197

Carnot cycle 36
CCL 190
Celsius 5
Characteristic functions 41
Characteristic point 150
Chemical potential 55, 60

Clapeyron diagram 99
Clausius-Clapeyron equation 65
Clausius' non-compensated heat 42
Cloud
   effect of freezing in 144
   enthalpy 78
   entropy 78
   internal energy 78
Components of a system 53
Condensation
   by adiabatic ascent 138
   by isobaric cooling 120
   by isobaric mixing 129
   nuclei 119
   trails 130
Conservative properties 152
Constant lapse rate layer
   internal energy 230
   potential energy 230
   transformations 233
Convective condensation level 190

Dalton's law 11
Dew 120
Dew point 116
Diagrams 98, 109
   aerological 98
   equivalent 99
Diathermic walls 3
Differential, exact or total 18
Dynamic
   altitude, height 161
   metre 161

Emagram 104
Enthalpy 21, 23, 45
   in the atmosphere 228, 229, 235
Entrainment 195
Entropy 35, 45
   of mixing 48
Equilibrium 2
   conditions of 43
   internal in heterogeneous systems 58
   metastable 3, 43
   radiative 176
   stable 3

INDEX OF SUBJECTS

unstable 3
  with small droplets and crystals 87
Equipotential surfaces 160
Expansion
  adiabatic in atmosphere 136
  pseudoadiabatic 142
  reversible saturated adiabatic 141

Fog 120
  advection 120
  mixing 130
  radiation 120, 219
  steam 130
Free energy 40
Free enthalpy 40
Freezing in a cloud 144
Freezing of small crystals 96
Frost 120
Frost point 116
Fundamental equations 41, 57
Fundamental lines 98
  relative orientation 102
Functions
  characteristic 41
  thermodynamic 10

Gases
  equation of state 10
  ideal 10
  enthalpy of 26
  entropy 48
  internal energy of 26
  mixture of 11
  thermodynamic functions of 47
Geometric derivatives and differentials 163
Geopotential 159
  altitude, height 160
  field 156
  metre 161
Gibbs-Duhem equation 56
Gibbs function or Gibbs free energy 40
Gibbs' theorem 49
Gradient
  thermal 163
Gravity 156, 158

Hail stage 137, 144
Heat 17
  capacities 20, 21, 22, 49
  latent 20, 26
  of moist air 76
  of reaction 21
Height computation 171
Helmholtz function or Helmholtz free energy 40
Heterogeneous 2
Homogeneous 2

Humidity
  parameters 73, 151
  relative 75
  specific 73
  variables 73
Hydrostatic equation 159
  integration of 171

ICAO standard atmosphere 167
Ice nuclei 119, 138
Independent variables, number of 61
Instability
  absolute 186
  conditional 186, 188
  convective 197
  latent 190
  potential 197
Internal energy 16, 23, 46
  of atmosphere 227
  of layer with constant lapse rate 230
Isobaric cooling 116, 120
Isobaric expansion coefficient 46
Isobaric surfaces 162
Isolines or isopleths 98

Joule-Thomson effect 24

Kammerlingh-Onnes' equation 11
Kirchhoff equation 26
Kollsman number 169

Lapse rate 163
  dry adiabatic 180
  environment 183
  for atmospheric ascents 180
  moist adiabatic 180
  parcel 183
  saturated 180
Latent heat 20, 26, 68
Layer method 192
Layer with constant lapse rate 230, 233
LCL 141, 189
Level of free convection 190
Level of non-divergence (LND) 206
LFC 190
Lifting condensation level 141, 189
Limiting geopotential height 164
Littwin 224

Magnus' formula 68
Margules 231
MCL 149
Mean adiabat, method of 112
Mean isotherm, method of 111
Mechanical equivalent of heat 20
Mixing

adiabatic isobaric 127, 129
entropy 48
horizontal 127
vertical 147
Mixing condensation level 149
Mixing fog 130
Mixing ratio 74
  determination in diagrams 110
Modifications 9
Moist air 69
  adiabats 78
  enthalpy 78
  entropy 78
  heat capacities 77
  internal energy 78
Molar heat
  capacities 21
  of change of state 21
Mole fraction 12

Neuhoff diagram 104
Number of components 53

Orientation of lines in diagrams 102
Oscillations in stable layer 191
Osmotic pressure 91
Overturning of air masses 231

Parcel method 177
Partial molar properties 55
Partial specific properties 55
Phase rule 61
Phase transition equilibria for water 62
Point function 18
Poisson equations 29
Polytropic expansion 146
Polytropic processes 31, 146
Potential
  chemical 55, 60
  temperature 28
  thermodynamic 40
Potential energy
  of atmosphere 227
  available 235
  of layer with constant lapse rate 230
  total 235
Precipitation, maximum rate of 224
Pre-frontal cooling 208
Pressure altitude 169
Principles of thermodynamics
  first 16, 21
  second 35, 39
Process curve 98
Process derivatives and differentials 163
Processes 9
  adiabatic, ideal gases 28

polytropic 31
pseudoadiabatic 142
quasi-static 10
radiative 217
reversible 10, 35
reversible saturated 137, 141
thermodynamic 35
Properties 1
  extensive 1
  intensive 1
  partial molar 55
  partial specific 55
  specific 1
Pseudoadiabatic diagram 107
Pseudoadiabatic expansion 137
Pseudoadiabatic process 142
Psychrometric equation 126

Radiative processes 217
Rain stage 137
Raoult's law 93
Reference level 177
Reference state 17
Refsdal diagram 106
Reversibility 9
Reversible saturated adiabatic expansion 137, 141

Saturation
  by adiabatic ascent 138
  by isobaric cooling 120
  ratio 90
Scale height 166
Showalter stability index 190, 210
Skew emagram 105
Snow stage 137
Specific heat capacity 20
Stability
  absolute 186
  changes in dry air 201
  criteria 178, 185, 193
  parameters 208
  vertical 176, 201
Standard altitude 169
Standard geopotential meter 161
State
  equation of 10
  functions 10
  variables 10
Stüve diagram 107
Sublimation of small crystals 96
Supersaturation 90, 119
Surroundings 1
Systems 1
  closed 1
  heterogeneous 2, 53, 58, 59

homogeneous 2
   inhomogeneous 2
   isolated 1
   of units 6
   open 1, 57

Temperature
   absolute scale 5, 36
   adiabatic equivalent 151
   adiabatic wet-bulb 150
   and humidity parameters 151
   Celsius 5
   empirical scales 4
   equivalent (isobaric) 123
   ice-bulb (isobaric) 126
   Kelvin 5, 38
   parameters 151, 211
   potential 29
   pseudoequivalent 151
   pseudoequivalent potential 151
   pseudo wet-bulb 150
   pseudo wet-bulb potential 151
   saturation 139
   scales 3
   thermodynamic scale 38
   virtual 74
   virtual potential 78
   wet-bulb (isobaric) 123

Tephigram 99
Thermal gradients 163
Thermals 197
Thermodynamic equations of state 47
Thermodynamic functions of ideal gases 47
Thermodynamic potentials 40
Thermodynamic processes in the atmosphere 116
Thermodynamic scale of temperature 36
Thermodynamic surface for water substance 64
Thermometer 4
Total potential energy 235
Triple point 4, 61, 72

Units, systems of 6

Van der Waals' equation 11
Vapor pressure 62
   of small droplets 87
   of solution droplets 90
Variance 61
Virial coefficients 11, 14
Virtual change 43

Water vapor 69
Work of expansion 8